W9-ARE-218

The Fungus Fighters

Elizabeth Hazen (left) and Rachel Brown at the Albany laboratories of the New York State Department of Health in 1955. Their collaboration in research, fostered by the resources of the laboratories, resulted in the discovery of nystatin there five years earlier. (Photography Unit, Division of Laboratories and Research, New York State Department of Health)

The Fungus Fighters

Two Women Scientists and Their Discovery

RICHARD S. BALDWIN

With a Foreword by

GILBERT DALLDORF

CORNELL UNIVERSITY PRESS

Ithaca and London

First published 1981 by Cornell University Press.
Published in the United Kingdom by Cornell University Press Ltd.,
Ely House, 37 Dover Street, London W1X4HQ

International Standard Book Number 0-8014-1355-9
Library of Congress Catalog Card Number 80-69821
Printed in the United States of America
Librarians: Library of Congress cataloging information appears on the last page of the book.

IN MEMORY OF
ELIZABETH, RACHEL,
AND GILBERT

Contents

	Foreword *by Gilbert Dalldorf*	11
	Preface	17
1.	The Neglected Mycoses	23
2.	Microbiologist from Mississippi	36
3.	New England Chemist	46
4.	A Unique Institution	60
5.	Search and Discovery	71
6.	Laboratory to Marketplace	93
7.	After the Invention	104
8.	Evolution in Philanthropy	118
9.	Concentration in Mycology	132
10.	Spreading the Word	138
11.	Invention Repays Research	151
12.	Diversity of Support	162
13.	Nystatin in the News	174
14.	The Final Years	185
15.	A Continuing Challenge	193
	Reference Notes	201
	Index	205

Illustrations

Brown as a Mount Holyoke undergraduate 81
Hazen during college years in Mississippi 81
Hazen and Brown with nystatin formulas, Albany, 1958 82
Pasture where successful soil sample was dug 83
Two antibiotics found in Hazen-Brown cultures of soil microorganisms 84
Chemical structure of two antifungal drugs 84
Candida and *Aspergillus* species: disease-causing fungi 85
Hazen, Brown, and Dalldorf, 1955 86
Summer biology program at Vassar 87
Fungal growth on undersurface of tree leaf 87
Tree infected with Dutch elm disease 88
Nystatin sprayed on works of art in Florence 89
Dalldorf and Brown at Hazen memorial 90
Dalldorf and first Dalldorf Fellow 91
Rachel Brown, 1979 92

Foreword

Landmark discoveries in medicine are compounded of unusual aptitudes and attitudes, plus a measure of good fortune. They have surprising consequences and often bring a flood of secondary discoveries, for they open new paths in unpredictable directions. It was so with Alexander Fleming and his penicillin, and also true of two American scientists, microbiologist Elizabeth Hazen and chemist Rachel Brown, and their discovery in 1950 of nystatin, the first antifungal antibiotic that proved effective and safe in the treatment of human disease.

The clue that changed Fleming's career and medical history was a contaminating mold on one of his bacterial cultures that appeared to influence the growth of the bacteria. It was a rather common mold which had aroused the scientific curiosity of another researcher. His laboratory was directly above Fleming's, and spores of the mold had somehow drifted into Fleming's lab, probably through an open window.

Fleming was prompted by that little clue—and his alertness to unexpected antibacterial activities in surprising places—to speculate whether the mold might not be useful in controlling infection. It proved to be quite simple to demonstrate that this was indeed true, that the mold produced something that dramat-

[Gilbert Dalldorf, who died in 1979, wrote this Foreword shortly before the death of Rachel Brown.—R.S.B.]

ically destroyed a variety of bacteria. But he found it extremely difficult to extract the substance with the techniques and facilities available to him in 1928. After obtaining a tiny bit of the extract and testing it, he preserved the culture and went about his other work. It was not until World War II that others, taking advantage of newer technologies and techniques, succeeded in producing penicillin in quantity.

Fleming's reports, however, aroused immediate responses among physicians everywhere. That filtered extracts of a particular culture grown in specified fashion in proper nutrient media might yield substances that could miraculously overcome several lethal infections prompted many bacteriologists to brew their own. One, Jonathan Gluckman of Johannesburg, used his homemade penicillin with dramatic success when Winston Churchill visited the African high command and contracted lobar pneumonia. Among his cronies, Gluckman is remembered as the chap who saved Churchill's life and won the war.

Sparks from Fleming's great discovery fell in many places. One ignited a bonfire in New Jersey, in the mind of Selman Waksman. Waksman was a soil bacteriologist with little or no concern for man's infections. His interests centered around microbial life in soils and humus, and the agricultural significance of the microorganisms one found there. The exciting news of antibiotics suggested to Waksman that certain soil organisms he knew so well might share the wonderful faculty of inhibiting disease-producing bacteria and be useful, as was penicillin, in the field of medicine.

It was a fortunate decision and Waksman proceeded forthwith to discover a number of useful antibiotics. The first, actinomycin, was much too toxic to be used as an antibacterial agent and was long neglected. Twenty-five years later its toxicity was found to be so much greater for certain malignant cells—tumor cells—than for normal ones that with the dosage carefully adjusted, it could be most useful in the treatment of a particular kind of cancer. And then in 1943, Waksman's efforts were crowned with great success when streptomycin was found. It has the great virtue of suppressing the growth of the bacillus that causes tuberculosis. Streptomycin broke the back of tuberculosis.

A spark from Fleming's discovery started another blaze in the Albany laboratory of the New York State Department of Health, long famous for its understanding and skills in the control of infectious diseases. The laboratory was also known for its practice—unusual now but almost unheard of then—of recruiting women for positions of high responsibility. There the search for useful antibiotics focused on agents that might be useful in the control of fungus diseases, for which there were no known and wholly effective therapies. The search was led by a woman, Elizabeth Hazen, who was highly proficient in mycology as well as bacteriology. Prompted by Waksman's success, Hazen searched hundreds of soil samples and found in one bit of earth a streptomycete of a kind previously undescribed. It proved to have surprising antifungal activity, and Hazen then faced the same problem Fleming had: how to extract and chemically define the active substance.

She found an ideal partner in Rachel Brown, another of the women scientists Augustus Wadsworth, my predecessor as director of the state laboratory, had brought to the remarkable institution he created in Albany. Brown was an exceptionally talented, persevering organic chemist familiar with the extraction of active ingredients from bacterial cultures, a demanding procedure that was of immense assistance to other members of our scientific staff.

The Brown-Hazen collaboration that ensued brought forth in a surprisingly short time a characterization of the active principle in Hazen's streptomycete and evidence that it was highly active against a number of important fungi and only slightly toxic in experimental animals. They first called their agent "fungicidin," but, finding that word already in use for another preparation, renamed it "nystatin," for New York State.

At the laboratory, we soon learned that although nystatin had great promise as an antifungal drug, it also entailed a world of practical complications. We were swept into problems and issues quite remote from traditional laboratory affairs.

It quickly became evident that provision of sufficient nystatin to determine its clinical usefulness properly (it had not yet been tested on humans) demanded production methods quite alien to the Albany laboratory. Great tanks would be required, and engineering skills and manufacturing experience, all of which we lacked. Conditions were required that would provide substantial yields cheaply enough that the finished product could be marketed at an acceptable price. The laboratory-scale methods Brown and Hazen had used meant that each dose would have cost a small fortune.

It was unwelcome news. The Albany laboratory had no established patent policy, and we were all determined not to diminish its public service character. Nevertheless, a patent was clearly necessary. Without the protection of a patent, no manufacturer would hazard the costs of developing a bulk production technique for what was still an unestablished product.

We were saved, as always, by friends. One introduced me to Research Corporation, a nonprofit organization that administers inventions for educational and scientific institutions. Another contributed the services of his highly respected legal firm to advise us of our rights and responsibilities. In a short time we had made arrangements with Research Corporation for patenting the invention and licensing it to an experienced pharmaceutical firm.

There were substantial problems in mastering large-scale production, but after many disappointments and trials this was accomplished and nystatin was so efficiently manufactured that it could be sold for 20 cents a million units. The nutrient on which the streptomycete could most economically be grown proved to be lowly peanut meal. Once nystatin was freely available, its usefulness became more and more evident, and its role expanded in fascinating fashion.

The royalties paid by the pharmaceutical industry eventually amounted to many millions of dollars, all of which were plowed back to advance science and technology. Hazen and Brown, by their own wish, received no share of the royalties, although both could have become rich if they had chosen to profit from their work.

The inventors did, however, receive many awards, citations, and honorary degrees. One that was most unusual was given to both by the American Institute of Chemists, which broke with tradition and changed the bylaws governing its Chemical Pioneer Award to recognize Hazen, who was not a chemist but a microbiologist. Further, and most gratifying to Brown and Hazen—both examples as well as proponents of women seeking careers in science—was the fact that they were the first of their sex to receive that award.

The timing of the Chemical Pioneer Award made it the last that Hazen received during her lifetime. It was presented to Brown, who accepted for both because of the other's illness, on May 22, 1975. Hazen died on June 24. She had been advised of the award, of course, but a small "award celebration" planned in the nursing home where she was confined had to be canceled.

I was asked to comment on Elizabeth Hazen's career at a memorial meeting of the Medical Mycological Society of New York a few months following her death, and I concluded with: "She would have approved Francis Bacon's admonition, more than three hundred and fifty years ago, that since all are debtors to their professions, 'so ought they of duty to endeavor themselves by way of amends to be a help and ornament thereunto.' Few of us are able to repay our professional debts so generously and magnificently as did our friend from Mississippi."

Another of those few is our colleague from Massachusetts, Rachel Brown, who has equally reimbursed her profession. Now in her eighty-first year, she still is reading for *Chemical Abstracts*, as she has for the past fifty years; still is active in her church, where she taught Sunday school for forty years; still going to the library at the laboratory in Albany; still corresponding with and entertaining at her home the friends and fellow scientists she has known in all parts of the world.

As for me, I look back on those years as my most satisfying and productive. While my role in the invention and development of nystatin was peripheral, I feel honored to have worked with these two marvelous women at that unique institution and to have had some part in what has proved to be another of the landmark discoveries in modern medicine.

GILBERT DALLDORF

Oxford, Maryland

Preface

This book was written to record and preserve, while some of the principals were still alive, a little-known chapter in American medical science that could have been overlooked. It is the story of two women scientists who saw the threat of the fungus diseases, brought forth a new life-saving drug, and used their earnings from it to fight the diseases and help other scientists with their research.

While the book centers largely on the fungus diseases, it also touches upon other work in the biological-medical sciences aimed at generating better knowledge of human disease processes and teaching more people how to approach some of the unsolved problems. It points up the beauties—and the frustrations—of research as a means of opening new horizons in science and as a powerful tool for fostering creativity in young scientists.

This account should serve to alert physicians and laboratory technologists to the fact that the fungus infections, which probably received scant attention during their schooling, warrant a closer look today. It may also strike a spark of inspiration in scientists looking to turn their talents to a difficult and challenging field of research. It should show that neither sex nor age need be a deterrent to those who have the determination to explore new frontiers. And it gives a rare example of selflessness in science.

Elizabeth Hazen, highly trained and long experienced in microbiology, was fifty-nine when she began to study mycology so she could help physicians diagnose fungus diseases. Learning to identify disease-causing fungi in patient specimens she examined in her laboratory, she was dismayed that there were no known antifungal drugs that were effective and not harmful. So she set out to find one.

Rachel Brown was an ascending star in her field of organic chemistry when at age fifty she was enlisted in Hazen's research. Together they brought the work to a triumphant conclusion with the discovery of nystatin. The discovery, and the way it was handled, not only brought to physicians an effective weapon for treating fungus infections, but for more than twenty years produced funds for a grants program supporting biological-medical research.

Gilbert Dalldorf, director of the Division of Laboratories and Research of the New York State Health Department, where Hazen and Brown did their research, was the unsung member of the team. He encouraged Hazen in her fungus investigation, brought her together with Brown for the chemical work, and backed their joint efforts with the full authority of his position. And it was he who arranged the administration of their invention so that it produced multimillion-dollar support for the researches of other scientists.

I had known Gilbert, Rachel, and Elizabeth for some fifteen years before I began the research for this book in 1978. Elizabeth had died three years earlier, but Rachel and Gilbert were very much alive, alert, and enthusiastic about joining in the project. After working closely with them for nearly two years, I was profoundly shocked when they died within weeks of each other at the turn of 1980. Only a month before, both had completed their reviews of this manuscript and Gilbert had furnished the Foreword.

Details of Elizabeth's earlier life were gleaned mainly from Clara Hazen, her "sister-cousin," from Conway Dickey and

Beulah Townsend, younger cousins, and from Sarah Humphries, a long-time friend. Rachel's brother Sumner, his wife, Ruth, and Dorothy Wakerley, Rachel's friend and companion for more than half a century, were major sources of information about her. Among them, they produced family archives, letters, clippings, photographs, and memorabilia that brought out facets of the two women's nonprofessional lives not known to many outside their immediate circles.

On the professional side, among the stalwarts of medical mycology who supplied scientific and medical data for this book, and who gave critical review, were Libero Ajello, Margarita Silva-Hutner, Michael Furcolow, Morris Gordon, Howard Larsh, and Norman Conant—whose deep concern about the fungus diseases was inspiring and contagious. Further contributions were made by other scientists cited in these pages who provided details of their research and checked the accuracy of the reporting. Currier McEwen, a philosopher-practitioner of medicine, gave the same broad-gauged judgment in his critique of the manuscript as he offered in his long service on the advisory committee of the Brown-Hazen grants program.

Special thanks are due Anna Sexton, former librarian of the Division of Laboratories and Research, whose fifty-year history was the source of much of the material on the institution at which Hazen and Brown did their work on nystatin. Dagmar Michalova, the present librarian, aided tremendously by making easily available the Division's reports and other publications.

At Research Corporation, the foundation that handled the nystatin invention and administered the grants program, I had full access to the records and wholehearted assistance from those who worked on the program. Particularly helpful were Charles Schauer, Sam Smith, Kendall King, James Coles, Willard Marcy, Hal Ramsey, and Steve Bacon. Sandi Kirshenblatt was the major contributor to manuscript preparation, and Jennie Ewanoski and Gweneth Gormley assisted in a multitude of tasks.

Sources of most quotations are provided in the Reference

Notes. Some material was taken from reports of the New York State Department of Health, Division of Laboratories and Research; from the grants files of Research Corporation; and from the private papers of Brown, Hazen, and Dalldorf, now in the archives of Research Corporation. In these cases, abbreviated references are made in the text. Much of the other information was given to me in correspondence and conversations.

RICHARD S. BALDWIN

Stamford, Connecticut

The Fungus Fighters

1. The Neglected Mycoses

The problem attacked in the 1940s by microbiologist Elizabeth Lee Hazen and chemist Rachel Brown was age-old but perceived by few. Training in diagnosing the fungus diseases—the mycoses—was meager; diagnostic tools were rather primitive; and specific antifungal drugs did not exist. Today, a generation later, there are many more researchers working in medical mycology and more physicians and laboratory technicians educated to look for and diagnose these infections. There are also vastly improved laboratory techniques, and there are now some effective antifungal preparations.

Yet there is still a surprising lack of knowledge and lack of interest—among scientists and physicians as well as laymen—in what has become an increasingly serious threat to human health. Ironically, the discovery and use of drugs that combat other illnesses have only heightened the danger of mycotic infections.

At the heart of the problem is the ubiquity of the fungi, most of which thrive in or on the earth, in soils and plants of all kinds, from which their spores escape into the air to be touched and breathed by living creatures. Since they are so pervasive, it is likely that practically everyone has been exposed at some time to one or more varieties of fungi capable of causing disease in humans.

Many of those exposed have developed one of the mycoses but have not been aware of it; in many instances their natural defenses produced antibodies to ward off the microorganisms and provide a degree of immunity against further infection from that type of fungus. Others whose defenses were not able to repel the pathogenic fungi contracted one of the milder forms of the diseases. They were not seriously ill, but were embarrassed, inconvenienced, or perhaps temporarily incapacitated by one of the dermatomycoses—fungus infections of the skin, such as athlete's foot or ringworm. Data from a National Health Survey of the U.S. Public Health Service show that in the early 1970s an estimated one-third of the civilian population of the country had some skin pathology—one or more significant skin conditions that should be evaluated by a physician.[1] The second most prevalent of these skin conditions were fungus diseases, which had been contracted by more than 8 percent of the total population. In the age group 30–65, the fungus diseases were the most prevalent; they were exceeded in younger people mainly by diseases of the sebaceous glands, particularly acne, and in older persons only by tumors, malignant and benign.

While these diseases of the skin affect millions of people, they are classified as superficial, are generally more of a nuisance than a danger, but are often extremely painful. They are, moreover, very costly. The medications prescribed for their relief account for a very large part of the more than $75 million spent annually for antifungal drugs. And they are not always minor. In some units of U.S. troops in Vietnam, 70 percent of the men had ringworm infections that sometimes covered almost their entire bodies. Aside from the severe discomfort caused by the infections, there were times when half of the men in these units were not available for duty. The fungus responsible was an extremely virulent dermatophyte strain carried by the rodents that infest the rice paddies. It was one to which the Vietnamese rarely succumbed, because they wore sandals and light, loose garments, while in contrast, the heavy protective clothing and boots worn by the GIs created the hot, moist environment ideal for the growth of the fungus.

Still other people—mainly farmers, gardeners, woodsmen, and horticulturists, have contracted a potentially more troublesome fungus disease called sporotrichosis. This infection causes nodular growths under the skin that may be self-contained or may become ulcerous, spreading to other parts of the body, but only rarely being a threat to life. Those working with soil, mulches, trees, or plants are most likely to contract the disease and they are most vulnerable if there has been an injury or break in the skin through which the fungus can enter. It can be contracted by others, however. At a construction site in Kansas where children had been playing on bales of hay, several developed skin ulcers on their arms, legs, and other parts of their bodies which persisted for weeks in spite of treatment with antibacterial drugs. A more accurate diagnosis identified the disease as sporotrichosis and led to antifungal therapy that cleared up the infections and averted more serious damage. The fungus was traced to the hay on which the children had been playing.

Unknown numbers of other people, regardless of occupation or location, have contracted one or another of the most serious of the fungus diseases—the deep, or systemic, mycoses—which can disable and kill. (Their number is unknown because these infections are subject to misdiagnosis, being mistaken for other more commonly known diseases; also, as is true of all fungus diseases, there is no requirement that they be reported to the Center for Disease Control, the federal agency that gathers figures for morbidity and mortality in the United States.)

In almost all cases, the deep mycoses result when airborne spores of the pathogenic fungi enter the respiratory tract, where they find a hospitable climate for growth. In areas where these diseases are prevalent, most people inhale the fungi without apparent effect, or suffer only minor respiratory illness. Some, however, are more susceptible and contract chronic diseases, harboring the microorganisms unknowingly until certain signs of distress appear, sometimes years later. These are symptoms that may or may not be correctly diagnosed, depending on the awareness of the physician and the capabilities of the local laboratory.

These potentially dangerous fungus diseases are usually found in rather broadly defined geographic regions, and the physicians in those areas are likely to be familiar with the ones found there. In this country, coccidioidomycosis occurs mostly in southwestern Texas, southern New Mexico and Arizona, and parts of California. Blastomycosis is prevalent in the North Central and Southeastern states. Histoplasmosis is endemic in the central Mississippi and Ohio valleys. Cryptococcosis, on the other hand, has not been pinned down to specific areas; according to one authority, it can be found in any region where there are laboratories competent enough to detect it.

A physician in the San Joaquin valley of California will be alert to the possibility of coccidioidomycosis in a patient having certain fairly common symptoms. In the primary stage of the disease, the symptoms will be seen as indicating a minor affliction—such as a mild influenza—not warranting medical care, as it will probably clear up by itself. In its progressive stage, however, the doctor will see the disease as a potential threat, with the possibility of becoming as malignant as cancer and leading to almost certain death. In New England, where coccidioidomycosis is seldom seen, a physician would not usually suspect it or detect it.

In Ohio, where histoplasmosis occurs frequently enough to be a suspected disease, accurate diagnosis in an instance several years ago may have averted some fatalities. Students and teachers at a junior high school had celebrated Earth Day by cleaning up the immediate environment, including the school courtyard and parking areas that had been a starling and blackbird roost. Within two weeks absenteeism had soared and dozens of students and teachers—not only those who had worked on the cleanup—had developed fevers and influenza-like symptoms which were interpreted as signs of histoplasmosis. Five people were affected seriously enough to require hospitalization, but none died. The health authorities knew that earth enriched by bird droppings is an excellent medium for nourishing the fungus responsible for the disease, and samples of soil taken from the site of the cleanup confirmed it as the source of the spores which caused the infection. The school's ventilation system had its air intakes in the courtyard; the system had spread the fungus spores throughout the school.

Some of the deep mycoses will respond to treatments that are available today and, like other infections, are more likely to be controlled if treated at an early stage. However, once the pathogenic fungi have gone out of control and entered the lymphatic system of the body, they can disseminate, spreading to other parts and producing infections that are all but impossible to arrest. Among the drugs most widely used against these diseases is amphotericin B, a powerful antifungal agent that produces severe side effects in some patients.

A complicating factor for physicians in diagnosis is that some of the deep mycoses mimic other diseases more usually encountered in general practice. Fungus diseases may cause nonspecific symptoms of a mild upper respiratory infection—such as a low-grade fever and cough, or chills, sweating, and headache. Further probing may reveal what appear to be symptoms of pneumonia, tuberculosis, meningitis, rheumatoid arthritis, brain tumors, or other afflictions. Unless the physician directs, and the laboratory performs, the highly specialized and time-consuming tests for pathogenic fungi, that possible cause of the disease may not be explored. The treatment then prescribed may be inappropriate, ineffective, and ultimately harmful to the patient in that it delays antifungal therapy that might have been effective at an earlier stage.

Candidiasis (also called moniliasis) and aspergillosis are fungus diseases that are more likely to be recognized and diagnosed accurately because they are now occurring more frequently in all parts of the country. The agents responsible are members of the genera *Candida* and *Aspergillus,* fungi which are among the commonest encountered and to which healthy individuals normally are immune. *Candida albicans* is present in the mouths and intestinal tracts of most people, but ordinarily it is not a threat because it is held in check by other microorganisms. Spores of the *Aspergillus* species are found in the air almost everywhere, even in operating rooms and isolation wards of hospitals, but usually are unable to make inroads against the immune system of a healthy person. Given a break in this defense, however, these "opportunistic" fungi—which are not innately pathogenic—take advantage of the lowered immunity to establish themselves at the most vulnerable sites.

Causes of lowered immunity may be diseases that result in the overall debilitation of the patient, or the effects of drugs administered in treatment of other diseases. Tuberculosis and cancer victims may be so generally weakened that they become unknowing hosts to *Aspergillus* or *Candida* species. Corticosteroid hormones used to combat rheumatoid arthritis may alter the immune system to the point that aspergillosis will set in as a secondary infection, perhaps even becoming the primary disease following remission of the disease being treated. Infections by *Candida* species may also occur in those who are being given broad-spectrum antibiotics for other diseases; the activity of the drugs destroys the bacteria in the intestinal tract that normally compete with fungi for nutrients, and with this competition removed the fungi proliferate rapidly. Also, the antibacterial drugs themselves depress the immunity of the patient. Immunosuppressants, given to improve the chances for success of organ transplants, are designed to lower the body's immune response so as to reduce the possibility of rejection, but they also make it more vulnerable to the opportunistic fungi. Surgery patients and those suffering from diabetes, severe burns, injuries, and drug addiction are also more susceptible to these fungus infections than healthy individuals. Women who are taking birth control pills are more likely to contract an opportunistic fungus disease than those who are not.

Nystatin, the Brown-Hazen discovery, is one of the most effective agents against opportunistic *Candida* and *Aspergillus* species infections in the intestinal and vaginal tracts and on the skin and mouth. Applied locally or taken orally, nystatin is brought into contact with the skin or mucous membrane surfaces where these infections have taken hold in patients whose natural defenses have been lowered. Combined with antibacterial antibiotics, the antifungal drug attacks the fungi in the intestinal tract that otherwise would take over when the competing bacteria are destroyed.

When fungi become secondary invaders in patients who are under treatment for other disorders, they make much more difficult the management of the primary disease and may themselves become threats to life. In a report to the National Institute of Allergy and Infectious Diseases, Michael L. Furcolow of the University of Kentucky College of Medicine cited these examples from five organ-transplant and burn centers:

> In one institution, 20 percent of deaths of patients with acute leukemia were caused by fungi.
>
> Fungus infections caused 27 percent of the deaths of burn patients in a university hospital and 35 percent of the deaths in a burn center.
>
> The cause of death of one-third of the kidney patients who died in a transplant center was a fungus disease.
>
> Of 100 liver transplant patients in a university medical center, 40 were infected by fungi, all but one case resulting in death. The agents responsible were *Candida* and *Aspergillus* species, the latter discovered not during the lives of the victims but only at autopsy.[2]

The liver transplant cases point up another difficulty in ascertaining the seriousness of the fungus diseases: unless one of the mycoses is suspected as a possible cause of death, the pathologist will not check for it and the autopsy report will show the cause of death as the primary disease that was being treated, not the fungus disease that may really have been responsible.

A reminder to physicians that they should be alert to fungus diseases was given recently by George E. Ehrlich of the Albert Einstein Medical Center in Philadelphia. He noted that in cases of suspected tuberculosis there are several fungus diseases—sporotrichosis, histoplasmosis, blastomycosis, and cryptococcosis—that should be considered in the differential diagnosis. They are less likely to be detected, will not be found unless specifically looked for, and may be identified only after failure of response to antituberculous therapy. "Since diagnosis can lead to containment and even cure," he wrote, "appropriate reminders of the ubiquity of fungi and the manifestations they produce will continue to be needed."[3]

Another physician familiar with the problem has made a similar comment, touching upon the role of public health authorities. Wilhelm F. Rosenblatt of the Tuberculosis Control Program, New Mexico Health and Environment Department, wrote to me in 1980: "Coccidioidomycosis and histoplasmosis of the lungs have often been mistaken for tuberculosis, and in times past patients with pulmonary mycoses have been admitted to tuberculosis hospitals where they, unfortunately, sometimes acquired tuberculosis on top of their fungus disease. For this reason and others, physicians should be trained to think of systemic fungus infections; in endemic areas public health could assume the task of reminding the medical community of the presence of this 'threat.'"

Such estimates as are available on the prevalence of fungus diseases and their importance as a public health problem are made from spotty and sporadic reports compiled on a voluntary basis by a relatively few epidemiologists and practitioners in the field of medical mycology. Libero Ajello, director of the Mycology Division, Laboratory Bureau, of the Center for Disease Control in Atlanta, has said: "Any attempt to quantitate the impact of the mycoses on public health is doomed to failure. Since they are not classified among the notifiable diseases, hard data on their incidence and prevalence, as well as information on the morbidity and mortality they cause, are either fragmentary or simply not available."[4]

In an attempt to make more generally available as much of this kind of information as could be obtained, CDC began in 1969 to gather, organize, and publish data voluntarily supplied by physicians and investigators around the country who maintained such records. Four years later this effort came to a halt when funds for CDC were cut, another casualty being the closing of its Kansas City field station which had been outstanding as a research and training center for medical mycology, and as a sponsor of similar programs in other institutions.

In his 1971 paper, Ajello also noted that two years earlier the Second National Conference on Histoplasmosis had passed a resolution recommending that steps be taken by CDC to have the mycoses classified as notifiable diseases, and that "the lengthy process for implementing this resolution has already been initiated." In 1980 the fungus diseases were not yet classified as notifiable diseases.

With the lack of full and authoritative data, any figures presented on the number of people infected by fungus diseases, the number moderately or seriously ill, or the number who die from this cause, are open to attack. Not only are the data fragmentary, but questions can be raised as to the accuracy of diagnosis, the uniformity of reporting techniques, the comparability of figures from different sources, and, particularly, any projections made from actual data gathered.

Nevertheless, it is useful to examine such data as are available if any sort of measure is to be applied to the seriousness of the fungus diseases and their public health implications. With these caveats in mind, the following data are presented as being as reliable as any that can be obtained.

In a paper published in 1974, members of the Ecological Investigations Program of the CDC, then located at the Kansas City station, reported data on the deep fungus infections, using figures gathered by the Commission on Professional and Hospital Activities (CPHA). The data they presented were for patients having a systemic mycosis as a diagnosis at discharge from those hospitals which reported voluntarily to CPHA. The individual reports were submitted by about one-third of the nonfederal, acute-care hospitals in the United States. This study showed that in 1970 a total of 2,192 patients having these fungus diseases as the primary or secondary diagnosis were treated in the reporting hospitals. Of this total, 1,304 patients had histoplasmosis, 416 had coccidioidomycosis, 107 systemic candidiasis, 105 actinomycosis, 95 aspergillosis, 77 cryptococcosis, 53 blastomycosis, and 35 sporotrichosis. (Actinomycosis has traditionally been included among the fungus diseases although the microorganisms causing the disease are now considered to be bacteria, not fungi.) Total deaths among these patients were 104, and the death rates ranged from 1:4 for cryptococcosis to 1:42 for coccidioidomycosis.

The authors of the study utilizing the CPHA data also calculated that 32,478 hospital days had been spent in that year by the 2,192 patients with systemic mycoses. Using a 1970 figure of $81 per day for hospitalization, they estimated the cost of treating these patients at over $2.6 million. Projecting these costs to the nation as a whole, they estimated that over $9 million had been spent in 1970 for hospital care of patients having deep mycotic diseases.[5] (A paper reporting similar data for 1976 is cited in Chapter 15.)

These data were obviously incomplete. Only hospitalized patients were included; those being treated by their own physicians for fungus diseases were not. Two-thirds of the nonfederal acute-care hospitals in the United States did not participate. Also excluded were the chronic-care hospitals and those operated by federal agencies, including the Veterans Administration.

Filling in part of the missing data, Furcolow in his 1976 report to the National Institute of Allergy and Infectious Diseases presented figures from the Veterans Hospitals for 1970, the same year as the CPHA study. A total of 1,299 patients were hospitalized with fungus diseases that year in Veterans Hospitals in the United States, the mycoses being given as the principal diagnosis in 599 cases and as the associated diagnosis in 700. Histoplasmosis accounted for 451 of the cases, candidiasis 318, coccidioidomycosis 236, cryptococcosis 75, aspergillosis 70, blastomycosis 68, actinomycosis 61, and sporotrichosis 20. Candidiasis was the principal diagnosis in only 34 of the 318 being treated for that disease, indicating that the others in that group had contracted it as a secondary infection, probably following treatment for other diseases.

The latest CDC figures available as this was written were those for 1978.[6] Since reporting of the fungus diseases to CDC is not required, any data submitted to it are voluntarily collected and interpreted by state epidemiologists in the individual states. CDC warns that these data should be used with great caution and in no case should be considered representative of the whole. It is fair to assume, however, that since not all states report, the actual totals for the United States are greater than those published by CDC.

CDC data on the number of acute cases of fungus diseases reported in 1978 show 1,212 cases of coccidioidomycosis in 12 states (1,096 in California alone), 771 histoplasmosis in 18 states, 55 cryptococcosis in 15 states, 53 blastomycosis in 11 states, 22 nocardiosis in four states, and 6 actinomycosis in four states. (Nocardiosis, like actinomycosis, is a bacterial disease that traditionally has fallen within medical mycology.) The 1978 total was 2,119 acute cases of fungus infections reported from 27 of the 50 states; there were undoubtedly others in the 23 states not reporting these diseases.

The total number of cases reported in 1978 was more than twice that of the previous year, swelled by a threefold increase in coccidioidomycosis and almost a doubling of histoplasmosis cases. The upsurge in these two diseases was most likely the result of epidemics in California and Indiana, respectively, reported here in Chapter 15.

Total deaths in the United States from acute fungus infections reported to CDC in 1977 were 688; of these, 237 were from moniliasis (candidiasis), 134 from cryptococcosis, 112 from aspergillosis, 58 from coccidioidomycosis, 55 from histoplasmosis, 18 from nocardiosis, 12 from actinomycosis, and 2 from blastomycosis; the remaining 60 were from unspecified fungus diseases. (Moniliasis is now more usually known as candidiasis, a more descriptive term since the disease is caused by species of *Candida* fungi.)

Deaths from candidiasis and aspergillosis, accounting for slightly more than half the total, have been growing steadily over the past ten years, more than doubling in that period. Aspergillosis deaths alone increased almost 100 percent from 1976 to 1977. These increases probably reflect the growing threat of the opportunistic fungi, which start as secondary invaders in patients with compromised immunity and may ultimately be the cause of death.

The total of 688 deaths from fungus infections reported to CDC for 1977 was considerably larger than total deaths reported for infectious hepatitis (508), and from two to five times as great as the totals for meningococcal infections (338), acute encephalitis (206), syphilis (196), and acute rheumatic fever (125)—all of which are notifiable diseases and presumably reported from all 50 states. On the basis of even this partial reporting of fungus diseases, they would seem to rank with some of the presently notifiable diseases as public health concerns.

Although there was good evidence in 1980 that the fungus diseases take a heavy toll in minor and serious illnesses and in fatalities, no national agency, private or governmental, was mounting a major attack on them, and there was little public interest to provide the groundswell that might induce substantive action. Even less attention, either professional or public, was being paid to them in 1944 when Elizabeth Hazen, already well established as a microbiologist, began to study medical mycology to help meet a growing need in laboratory diagnostic services.

2. *Microbiologist from Mississippi*

Elizabeth Lee Hazen was born on August 24, 1885, to parents she would barely have a chance to know. She was only two years old when her father, William Edgar Hazen, a cotton farmer, died at the age of twenty-nine. Her mother, Maggie Harper Hazen, was twenty-eight when she died the following year.

Lee Hazen, as she was known in her earlier years, and her older sister, Annis, were taken in by their maternal grandmother, Jane Welsh Harper. A younger brother, Willie Edgar, only a few months old when their mother died, and destined to live only four years, went to live with another family. A year later all three were moved again to become members of the family of their father's brother, Robert Henry Hazen; their ages almost matched those of three of his five daughters.

Lee's birthplace was Rich, Coahoma County, Mississippi, a tiny farming community a few miles east of the Mississippi River and some sixty miles south of Memphis. This was the county in which the southern branch of the Hazen family had been established before the Civil War. Lee's grandfather, Munson Hazen, a native of Vermont, had chosen to settle at nearby Friar's Point, following a journey down the Mississippi River on a raft. The Vermonter, then just seventeen years old, had evidently seen the promise of the land, which before the levee had been built was inhospitable, mosquito-infested, subject to flooding, and selling for less than fifty cents an acre. The 320-acre farm he bought near Rich, "The Hazen Place," had passed to Lee's parents the year she was born.

Robert Henry ("Lep") Hazen, a forester, and his wife, Laura Crawford Hazen, who reared Lee, lived in Lula, Mississippi, a few miles from Rich. Although Lee and Annis called them "Uncle Lep" and "Aunt Laura," the orphaned children were so tightly integrated into the family that the real daughters could not tell by their treatment that the other Hazens were not their own sisters.

The family was deeply religious and the children were reared strictly, each child becoming a member of the Lula Baptist Church, where they responded regularly to the church bell in the steeple and spent a goodly part of their younger lives in the front pew. Lee's interest in the church continued through her college years, but later waned; she did not transfer her membership as she moved from place to place in continuing her education and her career.

Robert Hazen had not gone to college, but he was determined that all of his children, including his nieces, would be afforded a strong educational foundation. He was a trustee of the Lula School and saw to it that it always had excellent teachers, one of whom boarded with his family for several years in the absence of other suitable accommodations in the small rural district. The Lula School that Lee attended, beginning in 1891, was an ungraded, one-room, one-teacher public school with some fifty pupils from six-year-olds to those of early high school age.

Just before her final year at the school, the family moved to a farm in adjoining Quitman County, but Lee stayed in Lula to live with her Aunt Laura's mother, Delilah Killian Crawford, while she finished school. The last year was spent in the new two-room, two-teacher Lula School, where her academic record was such that one of the teachers said that if she could give a grade of more than 100, it would go to Lee.

Even during these early years she was a keen student, letting nothing interfere with her schoolwork, spending less time in play than the others in the family, and having few close friends. She was a zealous reader, concentrating on history and biography, a preference that changed later as she stepped up her reading in science and medicine. Conway Dickey, a cousin who donated Lee's library to the Mississippi University for Women after her death in 1975, reported that there was not a single novel in the collection. Yet despite her serious reading, Lee had a sparkling sense of humor, often tempering her outspoken comments with flashes of fun.

She apparently decided at an early age to make a mark for herself. Clara Hazen, her "sister-cousin" and her contemporary in Robert Hazen's family, recalled the valedictory address Lee gave in 1904 as she completed her studies at the Lula School. The subject was Lucius Quintus Cincinnatus Lamar, a revered Mississippi statesman who served as representative, senator, secretary of the interior, and ultimately associate justice of the United States Supreme Court from 1888 until his death in 1893.

Clara Hazen believed that Lee's selection of the subject and research for the address provided a large part of the inspiration that led her to set her own sights so high. The actual delivery of the valedictory was probably the first time Lee took the public stage; in later years she delivered a number of scientific papers, but after the nystatin research she let the honor of reading jointly authored papers go to Rachel Brown.

Following her work at the Lula School, Lee went to Memphis, where she received private tutoring in the high school subjects that were not given in the rural school. In 1905 she traveled across the state to enroll in the Mississippi Industrial Institute and College at Columbus, founded in 1884 as the first state-supported college for women in the United States. (Later the institution became the Mississippi State College for Women and today is the Mississippi University for Women.) At the time she attended the college, tuition was free and the cost of boarding for Mississippi girls was less than ten dollars per month.

At the Institute she was entered in the preparatory department to finish the studies she had started at the ungraded school and continued with private coaching. After two years she went into the college proper, where she completed the usual four years of study in three, by taking an incredibly heavy schedule in the final year. Her interest in science flowered in college, her schedule including courses in physiology, botany, zoology, physics, plant physiology, and anatomy, as well as other academic studies. Her grades were only slightly better than average except in some of the science courses, in which she excelled.

In her last year the Scientific Industrial Course was introduced, she wrote, "to develop the practical side of the sciences, without in the least sacrificing complete and accurate knowledge of principles." This was perhaps a foreshadowing of the later work that would turn her knowledge of science to the most practical of ends—the alleviation of human suffering.

One of thirty-five graduating from the college on May 27, 1910, Lee was noted in the yearbook as having been secretary and treasurer of the Baptist Missionary Society and assistant business manager of the *Spectator,* the campus paper. She was also pictured as a member of the cast of an all-girl theatrical production, "Men and Maids from Gay Paris," and was the recipient of a Certificate in Dressmaking and the B.S. degree. "B.S. is rightly attached to her name, for she has a most scientific mind," the yearbook stated, noting that she intended to go to Leland Stanford University for advanced work. The account concluded with the expected but in her case accurate augury that "her quick repartee and good cheer win her many friends, who predict great things for her."

Her picture in the yearbook showed a sweet-faced, handsome young woman, immaculately coiffed and with a most serious look that typified her deeply held determination. It did not, however, hint at her flashing wit nor her ascerbic expression when confronted with opinions contrary to her own or thoughts she considered to be plain nonsense. There was also no indication of her height—she was just about five feet tall.

With her degree in hand, Lee moved that fall to Jackson, Mississippi, the state capital, where she taught physics and biology at Central High School for the next six years. During that tenure she pressed on with her education, but at schools different from her original choice. She attended summer sessions at the University of Tennessee where she studied biology in 1912, and at the University of Virginia where she took lecture and laboratory courses in physics in 1914.

After leaving her teaching post in Jackson in 1916, she went to New York to take up graduate studies in the Department of Biology at Columbia University, where she was told by one interviewer that either she had never been to college or that she had gone to a poor one. Nevertheless, she quickly proved herself and received her M.A. the following year. She then enrolled in a course in medical bacteriology at Columbia's College of Physicians and Surgeons, but this was cut short when she offered her services to the U.S. War Department during World War I.

By this time called by her new friends Elizabeth, instead of Lee, she worked as a technician in Army diagnostic laboratories in Camp Sheridan, Alabama, in 1918–1919 and in Camp Mills, New York, in 1919. Then, adding to her practical experience, she became assistant director of the Clinical and Bacteriological Laboratory of Cook Hospital in Fairmont, West Virginia, where she was responsible for the diagnostic and serological work of the hospital, a post she held until 1923.

Later that year, she left the hospital to return to New York and reenter Columbia for further graduate work in organic chemistry and to continue her study and research in the Department of Bacteriology and Immunology at the College of Physicians and Surgeons. From one of her research projects she produced in 1926 a report, titled, with her usual frankness, "Unsuccessful

Attempts to Cure or Prevent Tuberculosis in Guinea Pigs with Dreyer's Defatted Antigen."

The first of her papers to be published was her Ph.D. thesis on general and local immunity to ricin, a toxic substance occurring in the castor bean that had been subject to a great deal of study for many years.[1] Her thesis, a model of painstaking and scholarly literature research, contained a list of fifty-nine references to the previous work, along with concise summaries of the earlier findings. It was published in 1927, the year she received her Ph.D. in microbiology. Then forty-two years old, she was one of a relative handful of women who had received that degree from Columbia in the medical sciences.

Though this is largely conjecture, it may have been during this period that the only reported "romantic" episode of her life occurred. Although she admired men in their professional roles and maintained close friendships with some of her male colleagues, she never married. Always a very private person, particularly about her personal affairs, she did not confide such matters to even the closest members of her very close family. A cousin remembers some talk of Lee's engagement to a doctor unknown to the family, and that for some equally unknown reason they broke up. Perhaps she was by then doggedly determined to have her own career, or, as Clara Hazen put it, "You know, Lee was a bit bossy."

Shortly after getting her Ph.D., she met Sarah Burt (later Humphries), who was to become a lifelong friend, and established residence in the building that became her home for nearly half a century. Before her marriage, Sarah had lived with her aunt, Caroline Fielitz, in her apartment at 440 Riverside Drive, near the Columbia campus. When Sarah's room was vacated, Elizabeth rented it and later was able to afford her own apartment in the same building; however, she consistently used 3 Claremont Avenue as her address, although that street was actually at the rear of the building. The Claremont Avenue address seemed more "academic" to her because of its closer association with the university.

She continued her research at Columbia until 1928, when she was appointed resident bacteriologist at the college-affiliated Presbyterian Hospital, where she supervised the diagnostic work. She held that post for a year, leaving it when she became an active member of the teaching staff in the Department of Bacteriology and Immunology of the College of Physicians and Surgeons. This move got her back into the role she loved, one which would also prove to be valuable to her students in years to come.

Possibly the major turning point in Elizabeth's career came in 1931 when Augustus Wadsworth of the New York State Department of Health lured her away from Columbia with the offer of a job at its Division of Laboratories and Research. Following the indoctrination given all new staff members at the Albany headquarters, she was put in charge of the Bacterial Diagnosis Laboratory at the Division's branch laboratory in New York City, a location greatly to her liking, since she had by then put down firm roots in the city. More important, it allowed her to keep physically close to her beloved Columbia, since her apartment was only a few blocks from the university, and she was anxious to continue her association with it on one basis or another. As it turned out, she maintained that affiliation for another forty years, with Columbia playing a large part in the next major turn in her life.

In her application for the Civil Service examination required for the state job, Elizabeth slightly altered her vital statistics. Possibly because of the number of years it had taken for her to earn her way to the Ph.D., or possibly out of vanity, for which she was known, she was not always forthcoming as to her age. From early childhood, Clara Hazen recalled, Lee (Elizabeth) hated birthdays, especially her own. This was particularly odd in a family which made much of birthdays, always marking them as days of celebration when the more distant members of the clan were drawn together. Nevertheless, in her 1931 application she gave her birth date as 1888, three years off the true mark, and fudged the dates of her schooling to match.

The position at the branch laboratory was much to her liking and her performance there led to increasing responsibilities. Some years later, in what she called a "Biographical Sketch," she described her work in these terms: "I have direct supervision of work of a large number of technicians who are engaged in the examination of pathological specimens for the diagnosis of infectious diseases such as diphtheria, septic sore throat, typhoid fever, tuberculosis, gonorrhea, spinal meningitis, undulant fever, bacillary and amebic dysentery. In addition, I supervise the work of the serum diagnosis department in which the complement-fixation test for syphilis is performed on thousands of blood and spinal fluid specimens. In the absence of the associate pathologist who is not a full-time person I have the responsibility for the administration of the branch laboratory."

Elizabeth's major accomplishments there were in the field of bacterial diagnosis. She traced an outbreak of anthrax, a usually fatal disease in animals that is transmissible to man, to animal bristles used in a brush factory in nearby Westchester County. She was able to pinpoint sources of tularemia, another animal disease transmitted to humans and one totally unexpected in the New York area. And in another bit of scientific detective work, after the state laboratory had reported the first case in the United States of *Clostridium botulinum*, Type E, a cause of poisoning from improperly preserved foods, she traced the deadly toxin to imported fish—canned sprats from Germany and smoked salmon from Labrador.

In addition to carrying out her routine work at the laboratory and publishing research results on projects ranging from syphilis tests to diphtheria toxins, she renewed her association with Columbia in 1934 and 1935 when she attended lectures in organic chemistry in the evenings. A still more important connection with Columbia was made in 1944 when she took special work in mycology in the Department of Dermatology at the College of Physicians and Surgeons.

Augustus Wadsworth, then in his last year as director of the Division of Laboratories and Research of the New York Department of Public Health, had sensed the need for greater attention to be given to fungus diseases. The discovery of histoplasmosis in Panama had been followed by evidence of infection and outbreaks of the disease in the U.S. Middle West. Moreover, the growing use of broad-spectrum antibiotics had increased the number of patients overtly infected by pathogenic fungi.

These antibiotics, usually given orally, were extremely potent against a wide variety of bacteria, and thus cleaned the intestinal tract of the bacteria which in a normal, well-balanced environment kept the fungi under control. With the reduction in biological competition, fungus infections were free to run rampant. Laboratory identification of the pathogenic fungi became a growing and often troublesome problem, and better means of controlling fungus infections were needed.

In preparing to deal with the challenge, Wadsworth—having no mycologist on his staff—chose Elizabeth Hazen to become one. She had had no formal training in mycology, but was an experienced bacteriologist and an imaginative researcher. Further, her close association with Columbia provided an affiliation with the Mycology Laboratory at the College of Physicians and Surgeons, a distinguished research, teaching, and service laboratory for the study of fungi and fungus diseases. The Columbia unit had been established a quarter of a century earlier by J. Gardner Hopkins, under whom Rhoda Benham had become an authority on pathogenic fungi.

It was with Benham, a mentor she greatly admired and respected, that Elizabeth learned mycology. She started in the dermatology outpatient service in Vanderbilt Clinic, where she helped collect and examine specimens taken from patients, planting cultures at bedside and taking them to the laboratory for further observation and identification of fungal agents. Later she studied the more serious systemic fungi obtained from hospitalized patients or from Benham's stock culture collection. As Margarita Silva-Hutner, a colleague at Columbia, said later, Elizabeth "treasured every positive specimen, slide or culture, carefully preserving it, not only for her own use, but with Dr. Benham's encouragement, to take to her own laboratory, thus bringing mycological manna as well as gospel to her disciples and associates at the Division of Laboratories and Research."[2]

She also undertook as an independent investigation a project that would occupy her precious spare time for many years to come, a study of the fungus *Microsporum audouinii* that causes severe itching and loss of hair, mainly in children of school age. This research was put to work as she traveled to Dobbs Ferry, a short train ride up the Hudson from the city, where she volunteered her services to help children afflicted with tinea capitis, the disease for which *M. audouinii* is responsible.

Elizabeth lost no time in transferring her newly acquired knowledge of mycology and her own culture collection to the work of the Division. In 1944, shortly after she started her studies under Benham, she began to examine for evidence of mycotic infection batches of cultures sent to the branch from the Albany headquarters. Her collection of pathogenic fungi was put to use for her own further study and for training members of the staff in mycologic procedures. Reflecting her work, the annual report of the Division for 1945 noted that mycological consulting services were now being provided to the state-approved laboratories, and that special facilities had been made available to the physicians of the state for identification of fungi and related microorganisms.

Thus by 1945 Elizabeth's culture collection—later to become the basis for a major contribution to the literature of mycology—was well under way, as was her new career, that of medical mycologist. The ground was being prepared for her next major move—finding an antibiotic effective against fungus diseases.

3. New England Chemist

Elizabeth Hazen had passed her thirteenth birthday by November 23, 1898, when her colleague-to-be, Rachel Fuller Brown, was born in Springfield, Massachusetts. A thriving industrial and commercial center, Springfield was also within the sphere of influence of such eminent educational institutions as Mount Holyoke, Amherst, and Smith colleges, and the college that later became the University of Massachusetts. Fuller had been the maiden name of her mother, Annie Fuller Brown, but unlike Elizabeth Lee Hazen, Rachel rarely used either her middle name or its initial. Her father was George Hamilton Brown, whose real estate and insurance interests took the family, including Rachel's younger brother, Sumner Jerome, to Webster Groves, Missouri, in 1905.

During her elementary school years in the St. Louis suburb, Rachel's favorite subjects were drawing and painting, which she was advised to but did not pursue in later life, and—as a hobby—insects. The specimens she collected, mounted, and studied under the friendly guidance of a retired high school principal were far larger than the "bugs" that would subsequently engross her when she studied bacteriology, but the experience was probably her first hands-on encounter with one of the sciences, and one that stayed in her memory.

Rachel and her brother were trained early to be self-sufficient, the two alternating weekly between "boys' work"—taking out the ashes, making minor repairs, cleaning—and "girls' work"—dusting, making beds, helping with the cooking. Sumner thought his sister was the best cook in the world, a belief that dated back to still earlier days when he cheerfully ate the mud pies she made for him until their mother convinced him that they weren't for human consumption.

They also learned snatches of a foreign language at an early age, which they used between themselves. They had picked up words and phrases from a German family in their neighborhood, but didn't realize at the time that it was German or even that it was foreign. For years, Sumner's pet name for Rachel was "Mrs. Frick," the name of their German neighbor.

What had been a fairly normal childhood was interrupted in Rachel's last year of elementary school when their father left the family. Their mother then took over the wage-earning role and in 1912 moved Rachel and Sumner back to Springfield where aunts and uncles with whom Annie Brown had grown up still lived. They were shortly joined by Rachel's grandmother and ailing grandfather. It was a precarious existence, but for years Annie supported Rachel, Sumner, her parents, and herself—first as secretary of Christ Church in Springfield, then as director of religious education in other Episcopal churches in Massachusetts and later in Pennsylvania.

Back in Springfield, Rachel finished the last half-year of elementary school and entered Commercial High School—which her mother's aunts and uncles had urged as a fast way toward becoming self-supporting and contributing to the family. Her mother had disapproved of the idea in the first place, and after Rachel had completed the first semester, she could stand it no longer and transferred her daughter to Central High School, where she could get a "classical" education preparatory for college (the school has since been renamed Classical High School). Annie had been well educated at a private girls' school in Springfield, but had not gone to college, and was resolute that her children should have that advantage.

Rachel had had a taste of science at Commercial High School and did not particularly relish it, although civics and history fascinated her. At Central High she continued to be most interested in history and developed a liking for Greek but still had no feeling for science. She did, however, have a tremendous interest in Mount Holyoke College and made up her mind during the last year of high school that she would go there if it could be afforded. The college for women was nearby, and at that time tuition, room, and board amounted to about $425 a year.

She completed the four-year schedule at Central High in three and a half years by doubling up on courses, and on graduation in 1916 had a high standing in her class, the honor of a *magna cum laude,* and the reward of a $50 scholarship—which she considered a loan, not the true gift it was intended to be. An appeal to the Springfield Women's College Club for aid brought the response that no scholarship funds were available that year. But the college promised some financial aid, if needed.

Through an amazing stroke of fortune, it was not needed. What made Mount Holyoke not only possible, but actual, was the beneficence of a friend of Rachel's grandmother, Henrietta F. Dexter, whom Rachel called "Aunt Etta." Eugene A. Dexter ("Uncle Gene") had built a small bakery in Springfield into a major enterprise which he sold to one of the national companies, and he wholeheartedly endorsed his wife's charities. In Rachel's case, the largesse—immediately following near poverty—was almost incredible. Aunt Etta guaranteed four years of undergraduate work, plus clothes, other necessities, and spending money. That all this did not spoil Rachel was a tribute to her own integrity and the solid grounding she got from her mother.

Sumner Brown went to college also, but without the help of Aunt Etta, who felt that he should drop out of school at Central High and get a job so he could help support the family. Annie

Brown flatly refused to permit it, so he continued there while working after school at the local library. He graduated two years after Rachel and headed for college with fifteen dollars in his pocket and hope of financial aid.

The college he headed for, however, was not the one he attended. He had been accepted by Dartmouth and was on the train to Hanover, New Hampshire, when he met two students from Amherst who urged him to try their college. He got off at Amherst, Massachusetts, arriving on the campus without any previous contact. Within a few hours he was interviewed and accepted, promised a scholarship, and given a job at the college library and another waiting tables at a boarding house.

Overwhelmed that in one day he had been practically guaranteed a college education, Sumner was certain that this could not have happened without divine intercession, and at that moment decided to go into the ministry. He said later that he believed Rachel's career had been similarly guided—that she too had been "pushed by the hand of God" in her work.

At college Rachel followed her high school predilection for history, choosing it as one of her majors. In her second year she found that she would have to take either chemistry or physics as a science. Physics did not interest her, and chemistry not much, but she chose chemistry. Once into the subject, however, she said, "I was so thrilled by any of it that I thought I could spend the rest of my days just analyzing samples—silver, iron, or what have you." Looking back later, she was grateful that college in her day was less permissive, that it required study in the basic sciences.

While keeping up with her class work, she led a full extracurricular life as a member of the Bible and mission study group at the Young Women's Christian Association, and in several other societies. One was the Dramatic Club, but her sole public appearance as a performer seems to have been as a dancer in a senior class production. Her friends remembered her for her contagious schoolgirl giggle, which persisted into later life, she and a classmate from her home town being called "the Springfield gigglers."

Social life at the college in her day was quite proper. In her circle it was the custom for the young ladies to write rather formal notes of invitation to their fellow students for meetings, or to study, have tea, or make fudge. Rachel also went to plays and concerts at the college and in Springfield, and to dances at Amherst and other nearby colleges. With few exceptions, curfew was 9:30.

A casual sportswoman in college and for years after, she favored particularly swimming, hiking, and golf. She was almost held back from graduating, however, because she was laggard in the required physical education course and had not handed in her outdoor exercise card for the previous month. During the two years in which Sumner's and her own undergraduate days overlapped, her brother often visited her at Mount Holyoke and, being a hiking enthusiast also, in winter he occasionally made the twenty-mile round trip from Amherst on snowshoes.

Mount Holyoke had a long tradition in chemistry, dating back to Mary Lyon, who founded the college in 1837. Lyon had taken a course in chemistry at Amherst Academy, attended lectures in chemistry at the Rensselaer School in Troy and helped to furnish a chemistry laboratory at another school. When raising funds to start Mount Holyoke, she specified a chemical room with the same priority as recitation rooms, and after the college was in operation included chemistry as a required course, taught classes in the subject and performed experiments for the students.

Based in part on this foundation, Holyoke at the time of Rachel's attendance had achieved a reputation for excellence in science with concomitant excellence in science faculty, including Emma Perry Carr, who became one of the molders of Rachel's career. Knowing of her fast-mounting interest in chemistry, Carr urged Rachel to consider the University of Chicago—which she had attended—for graduate work. The advice was heeded, and after receiving her A.B. in chemistry and history at Mount Holyoke in 1920, Rachel prepared to move to Chicago, where she would continue to have some help from Aunt Etta for one more year.

In the interim she helped herself. Getting a summer job at the Fleischmann Laboratories in Peekskill, New York, she found practical application for her love of chemical analysis. She spent the time happily, doing routine analyses of incoming shipments of molasses and of the daily mash of growing yeast—her first and only experience in an industrial laboratory. Particularly refreshing were the late afternoons when Rachel and others of the staff savored a treat of the freshly baked bread and molasses, and swam in a pleasant cove of the Hudson River, an activity that in those days was not a threat to health.

Moving on to Chicago in the fall for graduate work, she majored in organic chemistry, getting her M.S. the following June. Her research project for the degree gave her excellent training but afforded little opportunity for exercise of the scientific imagination she would show later. It was a rather routine task of preparing barbiturate compounds, actually the continuation of an investigation the head of the department had undertaken for the armed forces.

Rachel had helped pay her way through the year at Chicago by working as a laboratory assistant; now she was determined to be completely on her own. She had no great interest in teaching, nor had she any formal training for it, but that appeared to be the obvious means of earning money. There was an opening that appealed to her at the Frances Shimer School near Chicago, a girls' preparatory school and junior college that had a close association with the university.

Taking the appointment at the school, Rachel found that she would be teaching not only college chemistry, but high school physics. The latter was the subject she had first avoided at Mount Holyoke, only to discover later that for her major in chemistry she would also have to take physics. Three years of teaching the two sciences to the young women at Frances Shimer convinced her that teaching was not the career she wanted and, after taking courses in chemistry, French, and German at Harvard in the summer of 1924, she went back to the University of Chicago for further graduate work in organic chemistry. Needing to choose a minor subject, she elected bacteriology. Her horizons were widening and she realized by that time, "I had to apply my chemistry to something; I just couldn't sit around all day analyzing."

Her savings from teaching and an assistant's job in the chemical laboratory were enough to cover most of her expenses at the university and to pay back the scholarship from Central High. The principal of the high school at the time of her graduation, William C. Hill, wrote her subsequently, recalling that she had repaid the scholarship, with 100 percent interest, to pass along to somebody else, "a unique incident in my experience." At Chicago, Rachel also received aid in the form of an Edith Barnard Fellowship, which, like the earlier award, she had no obligation to repay but later did.

One of her chemistry research projects for the Ph.D. involved the molecular rearrangement of a tricky compound called *para*-phenylbenzbromamide. Unknowingly, she managed a "spontaneous" rearrangement when she dried the substance on filter paper and put it on a balance to weigh it in preparation for analysis. The compound literally went up in smoke. Her concern was not for her own safety, but only for the balance, which she was happy to see had not been damaged.

Purely by chance, another of her research projects at Chicago, her first in bacteriology, put her squarely into a field in which she later won considerable renown. The project assigned to her was the repetition of research recently reported in the literature—the isolation of a pneumococcal specific polysaccharide, a substance that identified one of the types of bacteria causing pneumonia. This development led later to a method of standardizing antisera for treatment of the disease. The laboratory exercise was one that would prove to be of major importance in her later work.

In 1926 Rachel successfully completed her research projects and course work and submitted her Ph.D. thesis, but there was a delay in arranging her oral examination. Her savings used up and the delay continuing, she could wait no longer before seeking employment, so she applied to the Division of Laboratories and Research in Albany, New York. The choice was made on the recommendation of a friend, Lucena K. Robinson, who had taken her master's at Mount Holyoke while Rachel was an undergraduate. Robinson had also gone on to Chicago for further graduate work, later becoming a chemist at the Albany headquarters of the Division. She had reported enthusiastically on the training and the opportunities for research there, and Rachel rejoiced when her application was accepted later in 1926.

Establishing herself in Albany, she became acquainted with a young woman of about her age who would become her lifetime friend and companion—Dorothy Wakerley, office manager and later assistant general agent of a national insurance company. They first met at the women's group of St. Peter's Episcopal Church and on hikes of the Adirondack Mountain Club, activities in which both were deeply involved. Their professional lives were totally disparate, but in addition to their common interests they proved to be eminently compatible, and their friendship became strong and durable.

Three years after taking the job at the Division, Rachel went to live in the home where Dorothy had lived for years with a family that had practically adopted her after she arrived in Albany to go to college. Dorothy had four brothers, but since childhood had wished for a sister and in Rachel she found one. Eventually they pooled their resources and bought a house in the city which was large enough to accommodate Rachel's mother and grandmother and would later become a haven for a number of others at various times.

In her new post as research chemist in the Department of Health, Rachel soon became engrossed in her work and stopped pondering the administrative slip-up that had held up the granting of her Ph.D. Despite her doctoral education, albeit minus the degree, she was put through the same initial paces as other newcomers to the Division, working first in the media department.

While it provided a routine, expected service, the media department was basic to the whole operation, for it was responsible for accurate formulation of media—the many nutrient mixtures used to grow microorganisms for research, for diagnostic tests, and for therapeutic purposes. It also ensured freedom from contamination of the glassware and other supplies used in all the laboratories. There Rachel was indoctrinated with Director Augustus Wadsworth's carefully detailed instructions for all Division employees. The objective was to impress upon the staff the necessity for following precisely the approved procedures so as to safeguard the health of the public served by the laboratories. With life-and-death decisions sometimes hinging on the accuracy of laboratory diagnosis of disease, and with the potential for endangering large populations through contaminated or improperly prepared medications, there was no margin for error.

Following this introduction, which gave her a bird's-eye view of the various activities of the Division, Rachel was prepared to move into its scientific investigations. Her first assignment was to work with Elizabeth J. and Frank Maltaner, who were well along in a project to improve and standardize the quantitative complement-fixation test for syphilis. While more reliable than the original Wassermann test, these procedures varied from laboratory to laboratory. Moreover, they were still relatively crude and inexact, with the test determination of positive or negative not being conclusive. One of the reasons for the unreliability was the lack of knowledge about—and therefore the lack of control over—the active principle in the antigens used in the test, these being the substances that stimulate the production of antibodies which identify the disease-causing microorganism.

Rachel worked on the antigen research only briefly, however, then turned the project over to Mary C. Pangborn, a Ph.D. chemist from Yale who was a member of the scientific staff. Several years later, Pangborn succeeded in isolating the active substance, which she named cardiolipin. This discovery made possible the standardization of the antigen for the Maltaners' complement-fixation test, and a vast improvement in its reliability. Cardiolipin was also utilized by Rachel some time later when she developed a simpler precipitation test for syphilis.

The work she had done on pneumococcal polysaccharides at Chicago proved to be directly helpful when she was assigned to her next project in Albany. This plunged her into the midst of one of the two major Division activities for which it was already world-famous—the laboratory identification of disease-causing agents and of antibodies in patient specimens, and the preparation and standardization of antisera and vaccines. As the result of earlier research and experimentation at the state laboratories, both therapeutic and preventive preparations effective against whooping cough, dysentery, typhoid, and certain types of meningitis and pneumonia had been developed and furnished, often in huge quantities, for civilian use and for the armed services.

Pneumonia was still a public health threat in the late 1920s—before the discovery of the sulfa drugs, penicillin, and the later broad-spectrum antibiotics—and the best available treatment was the use of antisera. The disease was caused by not one, but many types of pneumococci, each demanding a specific antiserum. This is where Rachel's polysaccharide work fitted in, for this sticky carbohydrate substance (the "slime layer") encasing the pneumonia-causing bacteria contained the antigens which were type-specific, distinguishing the type of pneumococcus in the sample. The isolated polysaccharides provided means for a simple precipitation test for standardizing antisera used in treatment of pneumonia, and, later, the preparation of vaccines for its prevention. In a series of studies over the next fifteen years, Rachel extracted the specific polysaccharides identifying dozens of types of pneumococci and achieved an extremely high degree of purification of the antigens.

By 1933, Rachel's status at the Division was such that she was called upon to represent it at a scientific meeting in Chicago. Prior to her departure, Wadsworth, well aware of her long-pending Ph.D. problem, got in touch with Julius Stieglitz, who had been her thesis adviser, alerting him to Rachel's impending visit. On arrival in Chicago she visited her adviser—as illustrious in his field of chemistry as his brother Alfred was in photography—and learned that if she could spend a week there, he could arrange for her oral examination. With Wadsworth's enthusiastic approval, she extended her stay, took the orals on the work on which she was now an authority and passed with flying colors. After seven years in limbo she received her Ph.D. in organic chemistry and bacteriology from the University of Chicago. She was also elected to Sigma Xi, the scientific research society of North America.

Now entitled to be called Dr. Brown, she continued her work on the polysaccharides and publication of the results of research in which she had been collaborating with Wadsworth. Her first paper to be published in a scientific journal had appeared in 1931, one of a series of six written with Wadsworth over the next twelve years on chemical and immunological studies of the pneumococci.[1]

Rachel's research on the polysaccharides led also to their use as antigens for vaccines, but few were produced; with what eventually turned out to be scores of types of pneumococci, it was impractical to prepare vaccines for each. One combination did find an important use many years later—a vaccine effective against several of the most common types of pneumonia, the one recommended today for protection of elderly and debilitated persons, who are the most susceptible to the disease.

Toward the end of her studies of the polysaccharides, Rachel returned to the work she had done earlier on improving the tests for syphilis. While the discovery of cardiolipin by Pangborn had made possible a more reliable complement-fixation test, the procedure was still extremely complex, lengthy, and expensive. Seeking a simpler one, Rachel developed a precipitation test using cardiolipin which she thought would be faster and less costly. Shortly she was able to show that the new test could play a useful part in the Division's examination of blood samples for evidence of syphilis. The opportunity for comparison with the complement-fixation test came when both procedures were demonstrated at a serology conference in Washington. Rachel, who was working alone, completed her tests by early afternoon, then went out sight-seeing. The three-member complement-fixation group, with identical samples to check, was still laboring over its tests late into the night.

The precipitation test proved not to be a substitute for the complement-fixation test, but a valuable screening procedure. It was subsequently refined to the point where it permitted screening some 500 to 600 blood samples received each day at the Division, making definite appraisals as to the 90 percent which showed as negative. This reduced considerably the more complex and time-consuming further testing needed to determine which of those not in the negative group were actually positive. For this final determination the now definitive quantitative complement-fixation test was used.

Through her familiarity with Pangborn's work on cardiolipin, Rachel also picked up some nonscientific information that became useful a few years later. Pangborn had been publishing her research results at each stage of her work so that other workers in the field could take advantage of each new set of findings. She had not been aware of, or was not alerted to, the effect the publications could have when it came to patenting the discovery. Patenting cardiolipin was considered necessary, not for any possible monetary returns it might bring, but as a method of controlling the quality of the material prepared in other laboratories. As it became evident that such control was needed over the extreme variations in formulations prepared by commercial laboratories, the need for a patent became more obvious.

There was no patent counsel at the Division, and there had been no similar case in the past to serve as guide. Gilbert Dalldorf, who had succeeded Wadsworth as director of the Division, then pressed for the best patent protection that could still be obtained. Failure to file a patent application within one year of publication of Pangborn's research results precluded getting a product patent in the United States, but it was still possible to get a process patent. This was issued to Pangborn in 1948 and she assigned it to the people of the State of New York, giving the Division the power to enforce quality control on the products manufactured by the commercial laboratories. Worrisome as it was, the experience was valuable for Dalldorf when he faced the problem of patenting nystatin a few years later. It made less of an impression on Rachel Brown, who, like Pangborn, was impelled to push on with further research.

The research that engaged Brown next was a collaboration with Myrtle Shaw of the scientific staff, who was studying soil microorganisms having antibacterial activity. Rachel's assignment was to apply her chemical expertise to isolate the active principle—the specific substance having the antibacterial properties. She succeeded and the antibiotic proved to be effective against bacteria, but, like so many other promising antibacterial agents, too toxic for human use.

The experience, however, turned out to be most pertinent when in 1948 Dalldorf walked into Rachel's laboratory with Elizabeth Hazen. Hazen's work with disease-causing fungi—and the lack of a drug specifically active against them—had inspired her to undertake research toward finding an antifungal, substance. In testing various soil microorganisms she had found some that had this kind of activity but needed a chemist to isolate the active material. Dalldorf felt that Brown's skill, developed in her research in antibacterial substances, would qualify her to help Hazen.

Thus the collaboration began: Hazen the microbiologist—slight, peppery, impatient, intensely active, and infinitely resourceful; Brown the organic chemist—sturdier and a few inches taller, solidly dependable, seemingly imperturbable, and quietly powerful when necessary. It was an odd combination, but one that worked.

4. A Unique Institution

The place at which Brown and Hazen collaborated on their research—the Division of Laboratories and Research in Albany—was itself a major factor in their discovery. At the time they joined forces it had a history of more than a third of a century of dedication to medicine and public service, and it afforded them the laboratory facilities, special equipment, trained co-workers, inspired leadership, and scientific-academic atmosphere, in which their work could flourish.

"The Division," with headquarters in Albany and a branch in New York City, was an institution unique in the United States, patterned to a degree after the interdisciplinary German *"Institut."* A major arm of the New York State Department of Health, it was a combination of scientific laboratory, educational institution, and operating public health service. Research—both pure and applied—was built into its being as well as its name. It was an organization respected by scientists and public health officials not only in this country, but around the world.

The reputation of the institution derived in part from the meticulously prepared *Standard Methods* of the Division, a compendium of precise, detailed directions for the technical duties of laboratory workers.[1] With the results of their work affecting directly the health, well-being, and even the lives of the citizenry they served, it was essential that they be subjected to the most exacting standards that could be devised. *Standard Methods,* used originally for training and monitoring the Division's own staff, later set the levels of performance that had to be met before state approval could be given to independent local laboratories, which numbered over 100 in 1927 when the first printed edition replaced the earlier typewritten version.

In the preface to that volume, Wadsworth, then director of the Division, noted that imposition of the standards had eliminated incompetent and unreliable laboratories throughout the state and improved the performance of the approved units. "The advancement of standards," he wrote, "started with the issuance of an approval after inspection, and has now reached the stage of advanced requirements as to the qualifications of the personnel, inspection of facilities and equipment, agreement as to methods of technical and office procedure, periodic testing of the results of the work—thus, complete supervision as to the fulfillment of minimum requirements of approved laboratories."

Wadsworth also pointed to the role played in the upholding of standards by the New York State Association of Public Health Laboratories, to which the approved laboratories belonged. In addition, the Association gave the Division a network of sources of information on local public health problems as well as a means of dissemination of new knowledge coming from headquarters.

As word spread as to the value of *Standard Methods,* requests for copies came in. Gilbert Dalldorf, Wadsworth's successor, wrote in the foreword to the third edition in 1947, "The important place which *Standard Methods* occupies is shown by the testimony of visitors and correspondents from many parts of the world. Older editions have been interleaved with translations in foreign languages and worn out in service. There is no better evidence of the usefulness of any book nor better justification for the present edition."

Education and training were high on the list of priorities at the Division. New staff members, many of them with college degrees, entered as trainees, were rotated through various departments and then assigned to the sections where their talents fitted best. They had access to Albany Medical College, Union College, Rensselaer Polytechnic Institute, the State University at Albany, and other institutions in the Albany area, for specialized courses, with tuition being paid and time off given to attend classes. If there were enough requests for certain courses, faculty members of the various institutions were brought in, their stipends being paid by the Department of Health. In some cases, the institutions gave degree credit for work done at the Division.

Among the many nonstaff members coming to Albany for training at the Division were laboratory technicians from the approved laboratories in the state, who learned mainly diagnostic techniques. Workers from other laboratories in the United States and abroad observed and practiced in the various departments and were taught approved laboratory procedures. Pathologists aspiring to practice in New York State could be accredited only after successfully passing the examinations administered by the Division. In cooperation with the New York State Association of Public Health Laboratories, courses were given in various parts of the state for laboratory technicians.

Staff members were encouraged to attend regional and national society meetings and to make reports on their return to the Division. Seminars in which the staff participated were held at regular intervals and served as training for presentations to be made later to broader audiences.

It was the academic atmosphere as much as the research facilities and health services that attracted candidates for staff positions, among them Rachel Brown and Elizabeth Hazen. The two women scientists had come there at different times and for different functions, and they gained there the years of laboratory and investigative experience on which their discovery would ultimately be based. Also, as a part of the unique institutional tradition, they had been allowed unusually broad freedom in pursuing their project. Perhaps most important to their work was their mentor, the director of the Division.

When Gilbert Dalldorf had gone to Albany in 1945 to take up his duties at the Division, he inherited an organization founded thirty-one years earlier but having roots going back to 1880. Wadsworth, its head from its establishment in 1914 until his retirement in 1945, had assembled the physical plant, the equipment, and the people, and had molded them into an institution whose influence went far beyond New York State.

The second state laboratory to be established in the United States, the Division had from the start the dual mission of finding better answers to medical problems and bringing them to the people. As Dalldorf wrote some years later in his foreword to the fifty-year history of the Division, "The compelling forces in the beginning were the unequaled opportunities bacteriology, immunology and pathology provided for the prevention and treatment of infectious diseases, and the recognition by government of its responsibility to make such blessings widely available."[2]

The progenitor of the Division of Laboratories and Research was the Sanitary Committee of the New York State Board of Health, which was set up shortly after the formation of the Board in 1880. Its chairman, Charles F. Chandler, had a profound influence on the new organization as he carried over to it the precedent he had set when he was Commissioner of Health for New York City: *experimental laboratory research is the basis for sound theory and practice in public health and sanitation.*

The first laboratory facilities to be made available to the new unit were those of the Bender Hygienic Laboratory in Albany, which was privately owned, although affiliated with Albany Medical College and built on land given by the city. Starting in 1890, it made bacteriological examinations of water supplies and analyses of specimens from cases of infectious diseases. A few years later the state set up the first facility of its own, the Antitoxin Laboratory. This unit was to manufacture and standardize tetanus, streptococcus, and diphtheria antitoxins, and further investigate antisera for treatment of tuberculosis, typhoid fever, and other diseases.

The Bender laboratory continued to work closely with the Antitoxin Laboratory and in 1910, when the state took over the work done by the private firm, the Bender group helped train the staff for the new organization, as well as handling occasional overloads from it. The public laboratory, by then known as the State Hygienic Laboratory, was the immediate predecessor of the Division of Laboratories and Research when the latter was formally established in 1914.

Preparing and standardizing antisera continued to be a major mission of the laboratory as it began operations as the Division. Following the discovery of a diphtheria antitoxin in 1890, wide attention had been given to the development of other antisera that might be effective against other diseases. As each serum was effective against only one specific type of infection, separate searches were made for antisera that might have value in the treatment of other major diseases.

The research activities of the Division were intensified as the search went on, and successes were reported in the production of a vaccine for whooping cough, and antisera for dysentery and certain types of meningitis and pneumonia. A serum for lobar pneumonia was used with particular effectiveness in an epidemic among U.S. troops stationed on the Mexican border in 1916. In the same year a typhoid vaccine was produced at the Division and furnished in massive quantities for vaccination of troops being mobilized in New York State. In 1928 and 1929 the Division's production was severely strained to meet the demands for its serum effective against meningitis, which was reaching epidemic proportions in New York City. Also produced at the state laboratories and used worldwide were an antitoxin for certain types of botulism, and antisera that aided in diagnosis of infectious diseases.

By the time of Dalldorf's accession in 1945, a virtual revolution had occurred to close the illustrious chapter of the Division's work with vaccines and antisera. The development of sulfa drugs had caused a reexamination of the need for the serum work, which was nevertheless continued because for some cases antisera still afforded the most effective treatment. However, with the advent of penicillin and later the broad-spectrum antibiotics, the need for further serum studies was greatly reduced. The new drugs were comparably effective, less expensive, much simpler to use, and they were available in good quality from commercial firms. The Division promptly turned the antitoxin and serum laboratories to other purposes.

Ironically, the success of the new drugs against bacteria brought about an increase in the incidence of fungus diseases—another area to which the Division would later turn its attention. Dalldorf had welcomed the new antibiotics, of course, but he was concerned that their effectiveness against bacterial diseases would breed complacence, since there was still much to be learned about diseases that did not respond to antibacterial drugs.

Among the new purposes to which the vaccine and serum laboratories were adapted was the study of virus diseases, a need which had been recognized for many years and the urgency of which had been pointed up by epidemics of influenza and poliomyelitis. However, other commitments of the Division and the lack of suitable facilities had delayed until about 1945 the start of a program of virology. Facilities were critical, for Wadsworth had had a fear of innocently introducing viruses into laboratories that produced sera for human use. The ending of serum production and the erection of proper isolation facilities made possible the undertaking of virus studies on a steadily expanding scale.

This development allowed Dalldorf to transfer to Albany the virus investigations he had begun earlier in Westchester County, New York. There he had set up a virus laboratory in which he had inaugurated a study of poliomyelitis, an immediate and pressing problem in the area served by the laboratory. As he wrote later in "Days of Our Years," his unpublished memoirs: "Poliomyelitis was a dreaded summer visitor in Grasslands Hospital during the 1930s when the isolation wards became collecting stations for paralyzed children from the entire county." He was among the many who turned their research efforts toward solutions of that problem.

Dalldorf's research at Grasslands had suggested that there were other, unsuspected viruses associated with polio, and on his arrival in Albany he extended these studies with Grace M. Sickles of the Division staff. In 1948 they announced discovery of a family of hitherto unknown viruses, which they named Coxsackie for the village in New York State in which there had occurred several years earlier an epidemic of the poliolike disease to which the viruses were related.

Responses to the publication of their findings revealed that such viruses are worldwide in distribution, and that various members of the group are responsible for a number of other diseases. One was designated a poliomyelitis virus in Russia and another causes Bornholm disease, or epidemic pleurodynia (the Devil's Grip). Some thirty years later a group of U.S. scientists announced that they had established a link between a Coxsackie virus and juvenile diabetes.

A further result of the freeing of the Division's staff and facilities from serum and antitoxin studies in 1945 was the start of another new activity. Pointing to opportunities in a more comprehensive study of antibiotics, Dalldorf wrote in his first annual report at the Division, "A determined effort to establish such studies properly will be made in 1946." He also noted that in 1945 special facilities had been made available for the identification of fungi and related microorganisms. Within four years these developments would lead to the discovery of the first safe antifungal antibiotic.

That the investigators who made this major discovery were women should have been no surprise to anyone familiar with the organization and philosophy of the Division at that time. Both had been recruited by Wadsworth—Rachel Brown in 1926; Elizabeth Hazen in 1931. Both, in accordance with Division practice, had been encouraged to follow their independent research interests.

This freedom, which permeated the organization, was abetted by the Director's scientific staff, made up largely of women, which continued under Dalldorf to initiate research aimed at solving new problems as they came up, passing the projects on to others on the staff to carry out on a routine basis. They assigned the projects anywhere in the Division where there were people qualified to do them, and welcomed suggestions or outlines for experimental work from service as well as research personnel. Both Brown and Hazen were members of this group.

A major asset of the Division was its remarkable library, which for many years was the responsibility of Anna M. Sexton. Beyond acquiring the books, journals, and other publications bearing on the work of the Division, the library provided a scanning, digesting, and informing service that staff members found invaluable in keeping abreast of developments in their areas of interest. It also provided reference and editorial services for staff members preparing papers for publication or delivery at scientific meetings.

Paying tribute to the library in 1962 as Sexton was about to retire, Victor N. Tompkins, Dalldorf's successor, said, "This Division is literally a group of laboratories clustered about a reference library early developed and continuously refined for its needs. The library has long been the intellectual center and in some ways a symbol of the collective aspirations of the staff."[3] After her retirement, Sexton called upon her long experience, first as Wadsworth's secretary and then in the library, to write the Division's definitive history.

Internal communications in the Division were vastly aided by the secretariat, a group of four or five women who were secretaries to the heads of the several operating sections. This was another of Wadsworth's creations, and he had successfully fought with Civil Service to create a new category of "secretary" to meet the high qualifications he set, believing that the caliber of support personnel should be as high as that of the scientists and technicians. The requirements for the job included a college degree and training in science.

Headed by the secretary to the director, the secretariat helped coordinate the routine and scientific activities of the Division. Members occasionally attended and reported on scientific meetings, and met monthly with the laboratory committee, composed of heads of the several labs, and at those and their own meetings exchanged information on the progress and problems of their respective sections. The free give-and-take, the easy exchange of information on research in progress, and the independence of action of staff members, all played subtle but important parts in the Brown-Hazen collaboration. Even more critical was Dalldorf's contribution, although he later told me that "once 'the girls' got started they didn't need advice from me or anyone else."

Pathologist, researcher, teacher, and administrator, Dalldorf had had unusual training and experience when he was summoned to the post of director of the Division. Born in Davenport, Iowa, in 1900, he had received his B.A. in 1921 at the University of Iowa, where he met Frances Barnhart, his wife-to-be. His M.D. was earned in 1924 at New York University and Bellevue Hospital Medical College, one year ahead of Frances and two ahead of Currier McEwen, who would also play a major role in his later life. In 1926 he took up a fellowship in pathology at the Pathologisches Institut, Albert-Ludwigs-Universität in Freiburg.

He had been determined to study at the German institution with Ludwig Aschoff, a distinguished leader in pathology. This became possible when he got a job as ship's surgeon on a transatlantic liner which took him to Bremen. Going on to Freiburg, he met and was interrogated by "the Geheimrat," and invited to join the staff. The problem of how to pay for the year was largely solved when as ship's surgeon on a later voyage he wrote and sold to *Liberty* magazine the story of an event in which he had played a part—the mid-ocean rescue of crew members of a storm-foundered freighter.

His first staff positions after he returned to the United States in 1927 were as pathologist at New York Hospital and instructor in pathology at Cornell Medical College, where he was most influenced by James Ewing. Three years later he went to Grasslands Hospital, where he served as pathologist, while carrying a reduced teaching load at Cornell for several additional years. The laboratories he designed at Grasslands were the forerunners of the Department of Laboratories and Research of Westchester County, which he later founded and directed until going to Albany. His admiration for the state Division of Laboratories and Research was such that he patterned the county unit after it, and also copied the name to accent research, as well as diagnostic laboratory services.

Among other studies, he was able at Grasslands to complete a comprehensive review of vitamin deficiency diseases he had begun at New York Hospital in collaboration with Walter H. Eddy of Teachers College, Columbia University, and which led to publication of a popular textbook.[4] He also inaugurated the work that led eventually to the discovery of the Coxsackie viruses, during which he made pioneer observations of what he termed "the sparing effect"—a phenomenon of great current interest. Finding that children with Coxsackie virus disease, a rather mild infection, seemed less prone to contract poliomyelitis, Dalldorf deduced that certain of the viruses caused human cells to produce a substance which inhibited the growth of other viruses. The substance, now identified as interferon, is being widely investigated today for use in cancer therapy.

Dalldorf continued his research on the Coxsackie viruses after he went to Albany in 1945, and gave courses in virus diseases in the College of Medicine at the University of Buffalo. In addition to carrying out his administrative duties, he taught at the Division, and at Albany Medical College, where he was designated professor of pathology and bacteriology.

The full role played by Dalldorf in the nystatin invention and subsequent developments has been known to few people. After bringing Brown and Hazen together for the collaboration on the nystatin work, he backed them to the fullest, then prodded them into putting their invention on record at a scientific meeting, thus possibly preventing their being scooped by their competitors in the race for an antifungal antibiotic. And it was Dalldorf who masterminded the arrangement that protected the invention, hastened the commercial production of the new drug, and ultimately made possible contributions of over $13 million for the support of scientific research.

5. *Search and Discovery*

The work that led to the discovery of nystatin had begun in 1944 when Hazen returned to Columbia to study medical mycology in her off-hours, then to put her new knowledge to practical use at the New York City branch of the Division. From her apartment on Claremont Avenue, she shuttled between the Division's branch on East 25th Street and the Mycology Laboratories on West 168th Street, maintaining a full work schedule at the branch and employing her own time at Columbia. It was a punishing routine, but the seemingly frail microbiologist from Mississippi kept it up for years despite recurring attacks of gastric ulcers which persisted for much of her adult life.

Shortly after starting her informal work at the university, she had begun a study of specimens submitted to the branch for laboratory aid in diagnosis of disease, checking for evidence of fungus infection. Her objective was to establish standard methods of examination for disease-causing fungi, and through the collection of fungus cultures she was starting, to assemble materials for comparative study. The collection, which she greatly enlarged over the years, provided a reference of positively identified fungi, with which the generally unrecognized disease-causing microorganisms could be compared. It also served as a valuable teaching tool as she began to pass on to other staff members her increasing expertise in medical mycology. Later it was to become the basis of her book, *Laboratory Identification of Pathogenic Fungi Simplified*, a standard reference in the field.[1]

In her 1945 report to the Division, Hazen noted that as physicians became familiar with the new mycology facilities, the service should be in greater demand, and that the approved laboratories of the state would probably make use of it. In the next year the number of specimens examined for fungus infections trebled and requests were received for help in diagnosis of specific types of infection, among them blastomycosis, coccidioidomycosis, cryptococcosis, histoplasmosis, and moniliasis (candidiasis).

By 1948 the number of specimens checked for evidence of mycological diseases again had doubled, the report for that year commenting that since facilities for making these examinations were not yet available in the local laboratories, the requests for aid in diagnosis were coming in from physicians in all parts of the state. Pleas for assistance in identifying fungi were also received from physicians as far away as Colorado and Louisiana.

Hazen's collection continued to grow, having already been put to use not only for training Division staff members, but for instruction of medical students from nearby hospitals. As word of the collection got around, transplants of fungus cultures were requested by and furnished to state-approved laboratories and to teaching institutions. Having so assiduously gathered cultures from other collections and from clinical cases, now she was able to share her wealth of material with others.

In her major assignment at the branch laboratory, Hazen continued to supervise the examination of specimens received from the associated laboratories. In her section of the 1949 annual report of the Division, she commented on the increase in number of cultures obtained from patient specimens by the local laboratories and sent to the branch for examination. She muted her usual directness for the official report, but still managed to let some of her feelings come through. Noting that the number of cultures received for examination had increased by some 60 percent, she added that "the work involved in the study of these cultures was not entirely justified by the results." The reason, she pointed out, was that nearly two-thirds of the cultures had been contaminated while in the local laboratories.

Reflecting somewhat wryly on the infant state of the art of mycological aid in diagnosis in the local laboratories, her report continued: "Attempts are being made to encourage submission of specimens directly, when a fungus infection is suspected, since it would not only afford the opportunity to study the specimens by direct microscopic examination, which often furnishes very important information, but also to make a more complete cultural examination than the other laboratories are probably equipped to do."

Hazen's published papers now began to show the direction of her future work. One, appearing in 1945, reported on her laboratory's examination of specimens for evidence of fungus infections, two in 1947 were on the fungus *M. audouinii*—which developed into her long-term research project—and one in 1948 which was her first on the search for antifungal agents. The paper on the antifungal search, appearing in *Mycologia,* was written with Albert Schatz, then doing antibiotic studies at the Division.

The *Mycologia* paper stated that in the research reported to date the most promising antifungal preparations had no widespread application because of toxicity or other undesirable pharmacologic properties. "Consequently," it added, "no antibiotic agent approaching the efficacy of penicillin and streptomycin against bacterial infections is available in fungus infections either of the superficial or the deep-seated type." The publication then reported the investigators' screening of soil samples in a search for microorganisms active against pathogenic fungi and noted that "attempts to isolate antifungal agents are now in progress."[2]

Hazen continued this research with Albrecht Weber, also of the Division staff, examining still more soil samples for microorganisms which were effective against disease-causing fungi, and isolating those possessing such activity. She had left for later solution the problem of identifying the particular substances within the cultures that possessed the antifungal activity. For this phase a chemist was needed, and it was at this point that Hazen went to the Albany laboratories of the Division to talk with Gilbert Dalldorf, who took her to Rachel Brown's laboratory. Brown was soon caught up in Hazen's project, receiving Dalldorf's blessing to take on the new collaborative assignment if she could still handle her other work.

Brown sandwiched the antifungal project with the antibacterial studies and the other tasks assigned to her—including in-service training of Division staff members—just as Hazen was attending to her other chores at the branch while she was doing her antifungal research. Dalldorf was not a harsh taskmaster, but he believed strongly that a researcher should have other jobs to do—jobs that might be routine, but provide a useful service. From his own experience he knew that in the course of doing research there were often obstacles that seemed to stop forward movement; if the researcher had other things to do, contributions could be made there while the investigator waited for fresh inspiration. The researcher with nothing else to do could easily become frustrated and fail to find a way around the impasse.

The collaboration of Brown and Hazen on the antifungal project began in 1948, getting off to a smooth start and throughout remaining personally as well as scientifically rewarding to both. It was, in addition, the beginning of a friendship that lasted for over a quarter of a century. Working in separate laboratories some 150 miles apart, they depended heavily on the U.S. mails to get their experimental materials back and forth. Their experience, unlikely to be duplicated today, was that almost invariably each received the next day the shipment sent by the other, and that the glass containers and contents they sent back and forth arrived undamaged. No shipment was ever lost.

At the beginning of the research, Hazen had chosen two

pathogenic fungi against which the antifungal properties of the experimental materials would be tested. One was *Cryptococcus neoformans,* the fungus responsible for cryptococcosis, a chronic disease affecting the lungs, skin, or other parts of the body, but particularly dangerous when attacking the central nervous system. The other test fungus was *Candida albicans,*the agent causing candidiasis (moniliasis), an infection of the mouth, skin, nails, vagina, bronchi, or lungs, and one increasingly seen as a serious systemic infection in patients being treated with broad-spectrum antibiotics. There were no known antifungal substances effective against either of the test microorganisms and safe for human use.

In her search for antifungal materials, Hazen concentrated on soils. As a microbiologist, she was well aware that soils are rich resources of microorganisms, and Selman A. Waksman had shown through his discovery of streptomycin in 1943 that some of the soil organisms had antibacterial properties. Hazen conjectured that in soils she might find other microorganisms that had antifungal activity. She gathered many samples herself and entreated her colleagues who were traveling to bring back soils from wherever they visited.

She eagerly opened each new soil sample arriving at the branch laboratory in New York City, carefully labeling it as to place of origin and name of donor. A tiny amount of each was mixed with a sterile saline solution and seeded on a nutrient base until any actinomycetes—the microorganisms most frequently having antifungal properties—had grown to the stage where they could be identified visually. The cultures appearing to be of the type sought were placed on another nutrient base, on which the test fungi were growing, and their action against the fungal growths observed. Those that stopped the growths were then grown in larger quantities in liquid nutrients, placed in mason jars, labeled, and shipped to Brown in Albany.

Hazen's work up to this point showed that somewhere in these still crude mixtures were contained substances produced by the actinomycetes which had the power to kill or at least to arrest the growth of the test fungi. It was Brown's job to try to find the particular chemical agents having this property, somehow separate them from the mass, and refine them so that Hazen could test them further. Here, Brown's training as an organic chemist and bacteriologist and her experience in extracting active antibiotic ingredients from bacterial cultures were brought into play.

Each mason jar received by Brown contained a whole culture, either "static" (a mixture of broth and a matted growth called the pellicle) or "shake" (a suspension of the growth in the broth). Her first job was to determine whether the active principle—the true antifungal substance—was in the pellicle, the broth, or both. To do this, she tried various solvents to find those that could separate this substance from inert materials and extract it for further processing. Determining the proper solvents proved to be a most tedious task, but Brown pursued it methodically and with endless patience, eventually finding the ones that would extract active material from the pellicle, the whole culture, or the broth. Samples of each of these extracts were sent to Hazen for further testing for antifungal activity. Along with each sample went a memo describing it and its preparation in detail.

Hazen's next round of tests consisted of trying each of these solutions against the two pathogenic fungi in the laboratory to establish first whether the active principles had indeed been extracted from the crude mixtures, and then to measure their relative potency. For those samples which showed antifungal activity she tested increasingly dilute solutions of Brown's materials against the fungi, reporting on the smallest amount of each that would stop fungal growth.

After preliminary tests, Brown and Hazen concentrated their research on the antifungal substances produced from just two of the several actinomycetes they had worked with. One was microorganism No. 47205, which had seemed promising in the earlier stages of Hazen's work with Schatz and Weber. The fractions it yielded proved to have good antifungal properties, but, like so many of the other substances tested, were toxic to mice. When this was established, Brown and Hazen abandoned their work on No. 47205.

Focusing now entirely on the other microorganism—No. 48240, one of Hazen's isolates—Brown extracted active material from the pellicle, the broth, and the whole culture. To her surprise, No. 48240 yielded two antifungal substances which she was able to extract and separate. One, which she obtained from the broth, proved to be active against only one of the two test fungi, *C. neoformans,* and this she labeled Fraction N for *neoformans.* The other, extracted from the pellicle, showed considerable activity against both *C. albicans* and *C. neoformans,* and was named Fraction AN (see illustration at the top of page 84).

While Hazen's laboratory tests proved that Fraction N was definitely effective against one of the two test fungi, subsequent animal tests showed it to be extremely toxic. The investigators then abandoned their work on Fraction N as a useful antibiotic, but Brown's curiosity led her to further chemical studies of this fraction which proved it to be identical to an existing antifungal agent known as cycloheximide, which was used to control fungus growths on golf course greens.

Centering now on Fraction AN of No. 48240, for which Hazen was supplying increasing amounts of pellicle, Brown confirmed her earlier finding that the fraction had only limited solubility in water and discovered that as she purified it further it became less and less soluble—a finding that was to prove of major importance later. As she attained purer and purer materials, those that showed the greatest potency against the test fungi were used by Hazen against actual fungus diseases induced by these pathogenic fungi in laboratory animals—mice and rabbits in this case. The animal experiments would show whether the antifungal activity seen in the laboratory was protective in living creatures, and—equally important—whether the substances were toxic in test animals and, presumably, in humans.

Brown and Hazen reported to the Division in 1949 that Fraction AN had proved effective in laboratory tests against not only the two test fungi, but fourteen others, both pathogenic and nonpathogenic. It was a particularly active antagonist of *C. neoformans* and *C. albicans* even in extremely dilute solutions—indicating its probable potency against these two common disease-causing fungi. Further, the preliminary animal tests had shown it to be only mildly toxic to mice.

Continued experimentation allowed the two investigators to report in 1950 definite proof that Fraction AN had antifungal properties in both laboratory and animal tests, and that it had little toxicity for mice and still less for rats and guinea pigs. And they confirmed their statement of a year earlier that its "known physical, chemical, and biologic properties distinguish it from any of the antibiotics reported in the literature." Fraction AN was a previously undiscovered, relatively nontoxic agent effective in animals against at least two of the disease-causing fungi. Although it had not been tested in humans, it might prove to be the safe antifungal antibiotic they had hoped to find.

Fraction AN of No. 48240 then was given the name "fungicidin," which later was changed to "nystatin." Hazen, having established that the actinomycete from which fungicidin was derived was a previously undiscovered microorganism, coined the appropriate scientific nomenclature for it—*Streptomyces noursei.* The first word indicated the genus of actinomycete; the second was taken from the name of the family on whose property the successfully yielding soil sample had been found. Ironically, of all the samples Hazen tested from compost, peat, and manure, as well as various forest, field, and garden soils, the one that proved out was a field soil she had dug with her own hands while vacationing with her friends, the Walter B. Nourses, on their farm near Warrenton, Virginia.

Dalldorf, having kept up with the progress of the research from his post in Albany, stepped in again to precipitate the first public announcement of the Brown-Hazen discovery, and possibly help the inventors reach their goal ahead of their competitors in the race for an antifungal drug. The National Academy of Sciences at that time held regional meetings and had scheduled one for Schenectady in October 1950. The program chairman, anxious to get the top scientific people in the area to deliver papers at the conference, asked Dalldorf for candidates. Realizing that the Brown-Hazen work was almost at the stage where it could be reported in a journal or at a scientific meeting, Dalldorf proposed his two staff members. After they were scheduled, he informed them.

Brown recalled that when Dalldorf told them that they were on the program they were dumbfounded. While the research was well under way and they were satisfied that they did indeed have a new substance with important antifungal properties, they had not fully written up their work—and what they had written certainly was not in form to present to one of the most respected of American scientific societies.

Under the spur, however, they gathered their notes, rechecked their laboratory records and prepared a paper. Whether from shyness or unwillingness to take credit for her work, Hazen refused to deliver the paper, so it was Brown who made the presentation. The word of the discovery caused no great stir at the meeting, but a science reporter of the *New York Times* saw its possibilities. As Brown was leaving the podium, he practically snatched her copy of the paper from her hands. Hazen intervened, telling the reporter, "Don't you dare say that it's fit for human use!" She then lectured him on the dangers of assuming that the drug could be used for humans when at that point it had been tested only in animals.

The reporter heeded her words and the story which appeared in the *Times* the next day was reasonably tempered, although Hazen did complain about some of the wording. Its content, however, created some excitement. Soon the inventors' and Dalldorf's phones were ringing, and representatives of the pharmaceutical houses began to arrive in Albany. This response came as a great surprise to Hazen and Brown, neither of whom had thought much beyond pursuit of the research. They realized at once, however, their obligation to foster the further development of their work. Recalling their thoughts at the time, Brown told me some years later: "If we had an antibiotic that was useful, we were responsible in some way to see that it was made available. But we were in no position to carry out the evaluation in humans, no position to even prepare enough material for the tests in the first place. Obviously, it had to be patented before any company would take it on, but how could we make sure they would carry through?"

Dalldorf felt his responsibility just as strongly, and was painfully conscious of the difficulties encountered a few years earlier in trying to control the cardiolipin invention. He knew, as the pressures built up, that they had on their hands not only an important invention, but a major problem.

Brown as a Mount Holyoke undergraduate (c. 1919). The college's tradition in science, a required course in chemistry, and an inspiring teacher combined to head her toward a career as an organic chemist.

Hazen during college years in Mississippi (c. 1908). Her schedule was heavy with courses from anatomy through physics to zoology, a broad base for her later specialization as a microbiologist and medical mycologist.

Hazen (left) and Brown with nystatin formulas, Albany (1958). Top left on the blackboard is the empirical formula for nystatin; below it is the tentative structural formula; curve at right is the ultraviolet spectrum of the antifungal drug. (Photography Unit, Division of Laboratories and Research, New York State Department of Health)

Pasture where successful soil sample was dug. Of hundreds of samples Hazen tested from all parts of the country, the one that eventually produced nystatin was one she dug herself while vacationing with friends on the Walter B. Nourse farm in Virginia. (Photograph Richard S. Baldwin)

Two antibiotics found in Hazen-Brown cultures of soil microorganisms. Cycloheximide (from the broth) was already in use as a fungicide on golf courses, and nystatin (from the matted growth) was the first effective antifungal drug safe for human patients. (Photography Unit, Division of Laboratories and Research, New York State Department of Health)

amphotericin B

nystatin

Chemical structure of two antifungal drugs. Minute differences account for major differences in usage. Amphotericin B is soluble and can be injected for treatment of deep mycoses; nystatin, less soluble, is given orally for intestinal fungus diseases or applied directly to affected areas.

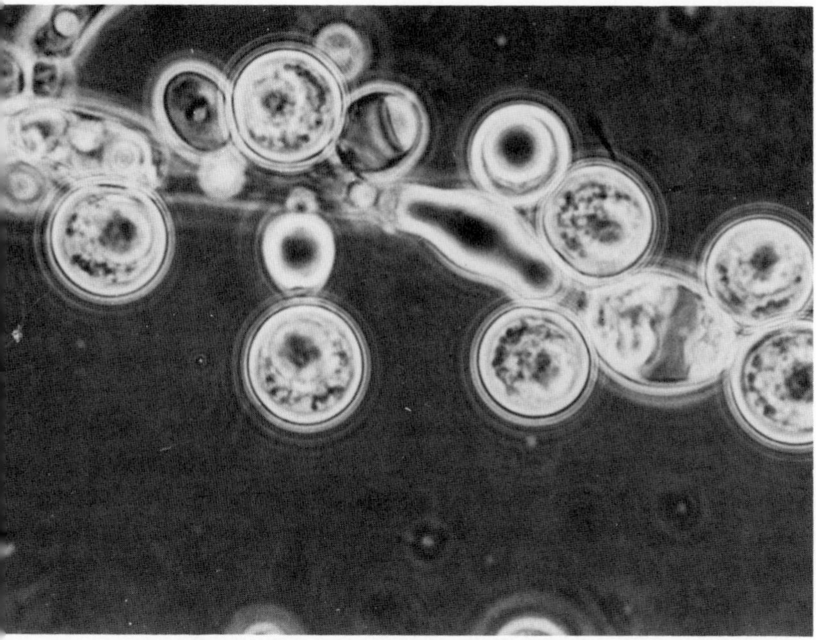

'andida and *Aspergillus* species. These re among the most widespread of the isease-causing fungi. *Candida albicans* :op), shown as a culture grown on ornmeal agar (X1125), is normally resent in the mouths and in intestinal nd vaginal tracts of healthy persons. *spergillus fumigatus* (below), shown s a direct examination of a patient pecimen (X750), is pervasive in the tmosphere, even in environmentally ontrolled laboratories. Nystatin is one f the most effective drugs used to ombat both these fungi. (Photomicro-raphs by Margarita Silva-Hutner)

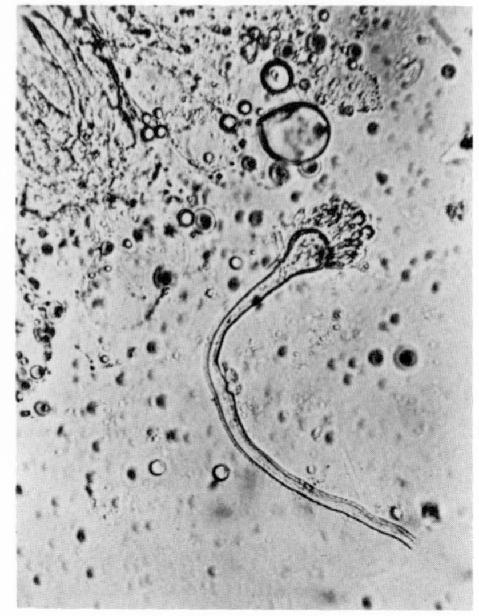

Hazen (left), Brown, and Dalldorf, 1955. They are shown here after the ceremony at which the two women scientists received the Squibb Award in Chemotherapy, their first major scientific prize. (Photography Unit, Division of Laboratories and Research, New York State Department of Health)

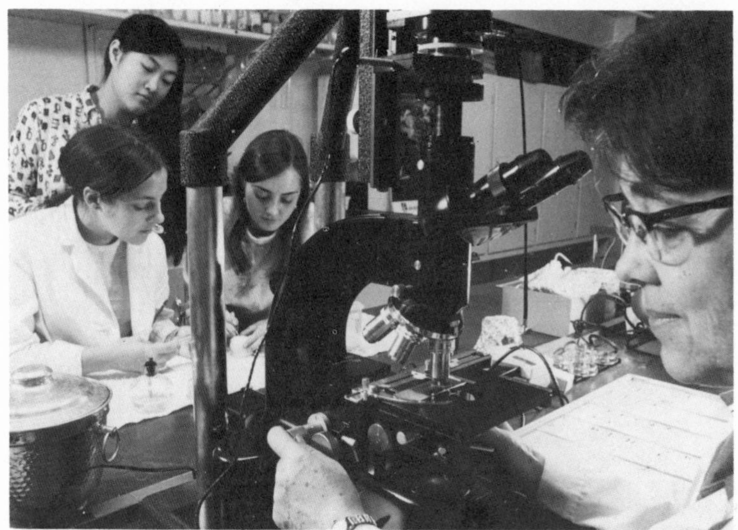

Summer biology program at Vassar (1969). Part of the Brown-Hazen Committee's effort to encourage women to take up careers in science, the program brought students in close contact with faculty on research projects. (Arax-Serjan Studios, Poughkeepsie, New York)

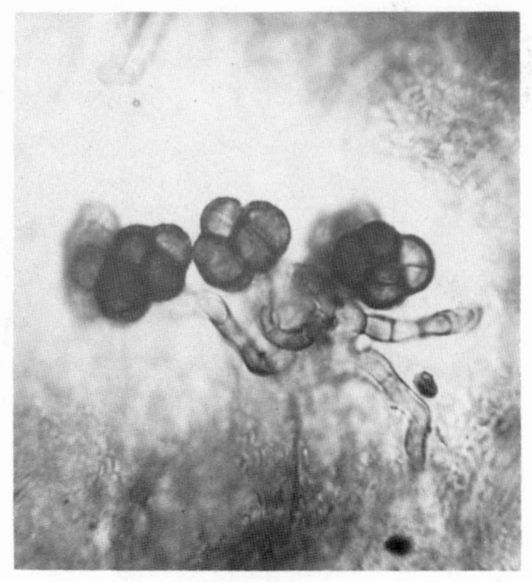

Fungal growth on undersurface of tree leaf. The plant pathogen (direct examination —X1200) causing Dutch elm disease responds to treatment with a solution of nystatin, which spreads to sites of infection through the tree's circulatory system. This use is an outgrowth of much earlier Hazen-Brown research on antifungal substances for human use. (Photomicrograph by Margarita Silva-Hutner)

Tree infected with Dutch elm disease. Tree is treated for fungus infection with nystatin mixture fed from tank on the trunk through a network of tiny tubes into woody tissues at the base. (Courtesy of Lowden, Inc., Needham, Massachusetts)

Nystatin sprayed on works of art in Florence. Masterpieces were saved from being devoured by fungi in the aftermath of the Arno River flooding in 1966. Art restorers found that the drug for humans would stop fungal growths on paintings without harming pigments. (Balthazar Korab © National Geographic Society)

Dalldorf and Brown at Hazen memorial, 1977. The laboratories are dedicated at Mississippi University for Women, where a Brown-Hazen grant for microbiology laboratory equipment honored their late colleague, a graduate of nearly seventy years earlier. (Mississippi University for Women)

Dalldorf (left) and the first Dalldorf Fellow, 1979. Thomas Kerkering, first recipient of the Dalldorf Fellowship in Medical Mycology, met the scientist for whom the fellowship was named when Dalldorf visited the Medical College of Virginia shortly before his death. (Research Corporation)

Rachel Brown, 1979. This photograph appeared with her last published work, an article in a special 1980 issue of *The Chemist* devoted to the role of women in science and the world. (Act One Studio, Albany, New York)

6. *Laboratory to Marketplace*

One of the first pharmaceutical company representatives to talk with Dalldorf about the Brown-Hazen invention was Geoffrey Rake of E. R. Squibb & Sons, later a division of Olin Mathieson Chemical Company. Rake was a trusted friend of Dalldorf's from his Westchester County days, when Rake had been on the staff of the Rockefeller Institute in New York City and frequently visited the county laboratory. Squibb had been a pioneer in the development of penicillin and, like others in the industry, was actively seeking new antibiotics. Rake read the story in the *Times* on October 18, 1950, and was on the telephone to Dalldorf that day.

In later discussions with Dalldorf and Brown, Rake spoke of Squibb's interest in the Brown-Hazen invention and in seeing whether it could be turned into a marketable product. He believed that it would probably cost his firm as much as $500,000 to find out whether a successful drug could be developed. Among other expensive items, it would have to provide a large-scale production facility to process materials in the huge quantities needed to produce even small amounts of a substance that had not yet been tested in humans. Squibb would not make this kind of investment unless it could have patent protection and exclusive rights to the invention for several years.

Dalldorf knew from his own experience with the pharmaceutical industry that no firm would undertake development without this kind of assurance. He was satisfied that with Squibb's resources it could carry the project through once it committed itself. Also, with its obvious interest it could probably be relied upon to push the work so as to have a product ready for the market at the earliest possible time—something the inventors and he wanted very much. But whether Squibb or some other firm was ultimately chosen, the first need was for patent protection.

The invention having been "disclosed" (in the patent sense) at the October 1950 meeting of the Academy, an application for a U.S. patent would have to be made before the first anniversary of that date. More serious, it was possible that someone else seeking an antifungal antibiotic could get a clue from the information divulged at that meeting and make prior application for a patent. This was not likely, for the culture from which fungicidin had been extracted was being carefully guarded, but it was a chance Dalldorf did not want to take. He knew of an instance involving a former colleague, Robert R. Williams, whose published papers on successive stages of his research on vitamin B_1 were the basis of successive patent claims made by a foreign pharmaceutical company, including one carrying a mistake not corrected by the inventor until much later. That case had dragged through litigation for years before finally being settled in Williams's favor.

Dalldorf's experience with the cardiolipin patent had convinced him that the normal channels available to him were inexpert and too slow, and that professional help was needed. He set out to find it, but it was through an unlikely chain of circumstances that the aid was enlisted.

Mary B. Kirkbride, who had been Wadsworth's deputy and long continued as a helpful friend to the Division, was familiar with the difficulties of patenting and licensing cardiolipin. Knowing that similar problems faced fungicidin, she consulted her brother, Franklin, a New York City financial adviser involved in a number of business enterprises and having deep interests in medicine and philanthropy. Franklin B. Kirkbride was also a close friend of Dalldorf's, and the two often had breakfast together at the Century Club in New York when Dalldorf was in the city for meetings.

It was at one such breakfast that Kirkbride suggested to his friend that he see a fellow member of the midtown club, one Joseph W. Barker, president of Research Corporation. This organization—despite its name—was not a commercial enterprise but a nonprofit establishment with expertise in patenting and licensing inventions for colleges, universities, and scientific institutions. Kirkbride thought that the handling of the fungicidin patent could safely be entrusted to it.

Classified as a private foundation, as were the Rockefeller Foundation and the Carnegie Corporation, Research Corporation provided one service unlike any offered by the others. It evaluated inventions arising as the result of research at educational and scientific institutions, then undertook the patenting and licensing of the inventions, as well as the obligation of defending the inventions against infringers or those contesting the validity of patents. In its other major activity, Research Corporation was more like conventional foundations in that it made grants for scholarly studies, its field of specialization being the support of scientific research. This aspect, too, was of interest to Dalldorf, for he saw the need for some sort of instrument to administer the funds from the fungicidin invention, if it should prove to be successful.

Once Dalldorf had met with Barker, tentative arrangements were made for Research Corporation to handle the invention. J. William Hinkley, director of the foundation's patent management division, was put in charge of the project, assisted by S. Blake Yates, who specialized in pharmaceutical inventions. Things began then to move rapidly.

First, it was necessary to prepare a patent application in the names of the inventors and to file it as quickly as possible. The application—a lengthy and extremely precise legal-scientific document—was drafted by a patent attorney retained by the foundation and assisted by its patent staff and by the inventors.

A legal agreement between the inventors and Research Corporation was also needed to set forth the responsibility of the foundation in patenting and licensing the invention and in receiving, allocating, and distributing any royalties that might result. This was drafted by the foundation's legal counsel in collaboration with the patent staff, the inventors, and Dalldorf. Dalldorf also sought independent legal counsel and required that there be a provision that the governor of the State of New York could cancel the agreement at any time "upon the ground that the public interest is served thereby."

To give Research Corporation the power to act before the U.S. Patent Office on behalf of the inventors, to license the invention, and to assure that the licensee used due diligence in developing it, an assignment of the inventors' rights to the foundation was needed. In this document the inventors would give their invention outright to Research Corporation for the philanthropic purposes stated in the agreement, reserving no financial returns for themselves—which was their express wish.

Further, and crucial to the development of the invention, there was a requirement for a license agreement between the foundation and the company selected as licensee. This was negotiated with Squibb, Research Corporation having agreed with Dalldorf that it was a firm with the financial resources and the interest in the invention to carry forward the development and to bring a product to the market at a reasonable price.

On January 10, 1951—only weeks after Dalldorf's first meeting with Barker—the agreement between the inventors and the foundation was signed. On February 1 the foundation filed with the U.S. Patent Office Hazen's and Brown's application for a patent on fungicidin. On February 21 the inventors assigned their rights in the invention to Research Corporation. On February 27 the license agreement between Squibb and the foundation was signed. It gave Squibb exclusive rights for five years to manufacture and sell fungicidin and the right to sublicense other companies to sell products containing the antifungal substance.

The agreement between the inventors and Research Corporation was patterned to a degree after the one entered into earlier by the foundation with Robert R. Williams, Robert E. Waterman, and Edwin R. Buchman, covering their inventions on vitamin B_1. In that case the foundation had applied for and obtained patents and licensed them exclusively to a pharmaceutical company for an initial period, with the firm bringing the new vitamin to the market quickly and at successively lower prices. Further, in the instance that Dalldorf was familiar with, Research Corporation had, in protracted litigation, successfully defended the Williams et al. patents against claims made by other inventors.

Royalties on the vitamin B_1 inventions were directed to the foundation's ongoing programs in science and technology, and to a new program to combat dietary diseases—the prime interest of the B_1 inventors. The Williams-Waterman program in human nutrition had generated important new research results and had funded a number of practical measures effective in combating infant malnutrition. It had, in addition, brought wide recognition and appreciation to the donors. A Brown-Hazen program of grants administered by the foundation could be expected to make like contributions in its field of science and suitably honor the donors.

The agreement with the inventors of fungicidin provided that, after deduction of certain minor specified expenses, Research Corporation was to devote one-half of all royalties received to its own philanthropic purposes and one-half to a special fund, later named the Brown-Hazen Fund. From the Brown-Hazen Fund, disbursements would be made to "nonprofit scientific and educational institutions and societies for the advancement and extension of technical and scientific investigation, research, and experimentation in the field of biologic and related sciences."

It was in February 1951, shortly after the signing of the license agreement, that Joseph Pagano, a mycologist on the Squibb scientific staff, visited Hazen's laboratory in New York City to receive transfers of the zealously guarded cultures of *Streptomyces noursei.* Hazen also registered a sample of the culture with the American Type Culture Collection, an internationally recognized organization which preserves microbial cultures having scientific and industrial interest. This was a necessary step in the patenting process, for the availability of authentic cultures from ATCC was needed to insure that the culture was indeed the one described in the patent application.

In March 1951 Dalldorf reported that Squibb was "brewing fungicidin" at its plant in New Brunswick, New Jersey, and in April the foundation was notified that the firm was experimenting with production in 800-gallon tanks, a step on the way to bulk production which later utilized 18,000-gallon tanks. It developed, however, that conversion of the laboratory-scale procedures used by the inventors into industrial-scale technology would be more difficult, take longer, and cost more than Squibb had expected. In addition to the huge tanks and special equipment needed for growing the *Streptomyces noursei* cultures in quantity, a number of nutrient broths were tried before one was found for optimal growth rate of the microorganism. Extraction and purification processes, modeled after Brown's but expanded to mass-production techniques, had to be perfected before enough fungicidin could be obtained even for experimental purposes.

Brown and Hazen were frequent visitors at the Squibb Institute for Medical Research in New Brunswick, where they met with scientists, engineers, and production people from Squibb's nearby manufacturing facilities. They consulted on the chemistry of the process, the growing and maintenance of cultures, and other matters in the transition phase, and, in general, watched over the invention, which, Hazen told one of her friends, "is my baby, and I'm going to look after it." Brown, always a gentle person and not given to pronouncements, showed her inner steel at one point when she felt that things were not moving fast enough. Her soft but firm "suggestions" brought a reassignment of personnel to the project and, she felt, a new impetus to the development work.

While the technology was being developed by Squibb, additional patent applications were filed by Research Corporation in five foreign countries—all of which patents were eventually granted. The U.S. patent, however, was being held up for lack of evidence of "utility," which could be proved only by conducting clinical tests in human subjects. It was 1953 before fungicidin could be made available in sufficient quantities for these tests. When positive results came in from the physicians using the drug experimentally, they were forwarded to the Patent Office as evidence of utility. They were also sent to the Food and Drug Administration for support of the claim of efficacy needed to obtain FDA approval before the drug could be marketed in the United States. With utility proven, further prosecution of the patent application continued, and in August 1954 FDA approval was given for sale of tablets of Mycostatin in oral dosage form. Mycostatin was Squibb's trademark for the Brown-Hazen drug, which by now also had a new generic name—nystatin. In the course of its patent search on the invention, the Research Corporation group had found that the name "fungicidin" was already in use for another material, and so advised the inventors. Brown and Hazen had then renamed their discovery "nystatin," which they pronounced "nye-*state*-in," for the New York State Division of Laboratories and Research.

In September 1954 Squibb announced the availability of Mycostatin in tablet form, describing the product as "the first broadly effective antifungal antibiotic available to the medical profession." It was recommended for the prevention and treatment of intestinal moniliasis, or candidiasis, its use being indicated for patients treated with oral antibacterial antibiotics, especially when treatment was intensive or protracted. It was also recommended for prevention of intestinal moniliasis in intestinal surgery.

The announcement explained that the intestinal flora of patients treated with oral antibiotics, particularly the broad-spectrum preparations, undergo profound changes, with attendant strong overgrowth of *Candida,* the extent of the fungal growth appearing to be proportional to the amount of antibiotic taken. Mycostatin was said to clear up established monilial infection of the gastrointestinal tract within 24 to 48 hours.

Shortly before Squibb had made its new product announcement there arose another obstacle to issuance of the patent on nystatin. An article in the *Journal of Investigative Dermatology* in 1954[1] had seemed to imply that a Brown and Hazen paper published in 1949 in *Proceedings of the Society of Experimental Biology and Medicine* had "disclosed" nystatin.[2] If this were true, more than one year would have elapsed between the date of that publication and that of the filing of the patent application, preventing the allowance of a U.S. patent.

This was naturally of great concern, particularly to Squibb. The earlier Patent Office objection based on the lack of "utility" had not been so much of a problem, for the firm was confident that the trials with humans would be successful—as indeed they proved to be. This, however, was a real threat to the hundreds of thousands of dollars Squibb had already invested on the assumption that the patent would ultimately be granted.

Research Corporation's position before the Patent Office was that the 1949 paper contained no implication of disclosure, that Brown and Hazen had described in that paper only one of the processes they had used in purifying and concentrating antifungal substances and had not disclosed in any way the invention claimed in the patent application. To document the case before the Patent Office, the foundation's patent counsel prepared an affidavit to be signed by the senior author of the 1954 article, stating that any such implication was incorrect. The author signed willingly, since he had had no intent to make the implication, and the affidavit was added to the accumulation of exhibits filed with the Patent Office.

The foundation continued its prosecution of the patent application through 1955 and 1956. Finally, on June 25, 1957, U.S. Patent No. 2,797,183, covering nystatin and its method of preparation, was issued—some six and a half years after the patent application had been filed. The delay, exasperating as it had been at the time, later proved to be a boon to scientists whose research was supported by grants made from the nystatin royalties. With Squibb having completed development while the application was pending, nystatin was on the market for the full seventeen years of patent coverage, and royalties were paid over the entire period. If a marketable product had not been available at the time the patent was granted, the term of royalty generation would have been effectively reduced.

Meanwhile, Squibb continued to increase its production of nystatin for its own sale and for other pharmaceutical firms to which it sold the material in bulk. Shortly after the announcement of the availability of Mycostatin, its name for nystatin alone, Squibb introduced Mysteclin, nystatin combined with tetracycline, an antibacterial antibiotic, and later other manufacturers also combined nystatin with tetracycline or other antibacterial drugs under various trade names. These combination drugs later were banned from the U.S. market for lack of proof of efficacy, but continued to be sold in other countries. Nystatin itself was not affected, being more and more widely used in this country and abroad.

With nystatin effective against *Candida* infections of the mouth, skin, and intestinal and vaginal surfaces, Squibb eventually produced Mycostatin in a number of forms suited to these applications: oral tablets and suspensions, creams, ointments, topical powders, and vaginal tablets. Other pharmaceutical companies sold nystatin primarily in combination with other formulations under such trade names as Achrostatin, Declostatin, Mycifradin, Tetrastatin, Tetrex, and Nysta-Cort and Nystaform.

With Squibb having more than fulfilled its commitments to develop nystatin for commercial use, to bring it to the market quickly and at a reasonable price, and to provide the material to other manufacturers, its exclusive license was extended to 1961. During this period, the companies buying nystatin in bulk from Squibb for sale in their own formulations did so under sublicenses, with royalties on their finished products being paid directly to Research Corporation at the same rate paid by Squibb.

Among the companies sublicensed were most of the other major drug houses—Upjohn Company, Chas. Pfizer & Co., Lederle Laboratories Division of American Cyanamid Co., Dome Chemicals, Inc., Merck & Co., Mead, Johnson & Co., Bristol Laboratories, Laboratorios Wyeth, Inc., Armour Pharmaceutical Co., and Schering Corporation. In 1961 the Squibb license became nonexclusive, and another nonexclusive license to manufacture and sell nystatin was negotiated with Lederle, putting it into direct competition with Squibb.

That nystatin was meeting an important medical need was shown by the increasing amounts of the drug being prescribed by physicians. Royalties paid on the invention, based on a percentage of sales, started modestly, with some $38,000 in 1955. They went to $100,000 in 1956, $200,000 in 1960, $500,000 in 1966, $1 million in 1969, and $2 million in 1974, the year the patent expired. A total of over $13.4 million was paid over the period 1955 through 1976, of which $6.7 million went into the Brown-Hazen Fund and an equal amount to the foundation.

The Research Corporation share was expended for support of basic research in the physical sciences conducted by faculty members at colleges and universities, and for services to these institutions in bringing their inventions from the laboratory to the marketplace, as had been done with nystatin. The Brown-Hazen Fund supported research and training in the medical-biological sciences, provided aid to women scientists in academic institutions, and, in the final years of the grants program, added its force in the fight against the fungus diseases. For several years the Brown-Hazen grants program was the largest single source of nonfederal funding for research and training in medical mycology in the United States.

7. *After the Invention*

The flurry of phone calls and visits from representatives of major pharmaceutical firms that followed the announcement of the Brown-Hazen discovery in 1950 eased off after the agreement was made with Squibb early the next year. However, requests for cultures of *Streptomyces noursei* continued to come in. Some of these Hazen honored directly by sending subcultures from her own collection and others she referred to the American Type Culture Collection, with which she had registered the microorganism. The cultures sent were ostensibly for the further research of other scientists, but some must have found their way into other hands; at least, this is what Brown and Hazen surmised some years later when they saw nystatin being made at a pharmaceutical plant in Czechoslovakia.

"Fan mail," as Hazen called it, came in from many quarters after the discovery, most of it requesting samples of nystatin for treatment of fungus diseases suffered by the senders. Much as they would have liked to comply, the inventors had no surplus supply, and even if they had had, they were not permitted to send samples for human use. Prior to 1954 the new drug did not have FDA approval, and after its clearance it was available only with a physician's prescription.

Overall, however, the discovery had not made much difference in either the personal or working lives of Brown and

Hazen. Both continued with their assignments at the Division, Hazen at the New York City branch laboratory, Brown in Albany.

Hazen was an active worker in, as well as the supervisor of, the Mycology Laboratory, checking for pathogenic fungi in the patient specimens sent in by physicians and other laboratories. Her collaboration with Brown was maintained, as they did further research on nystatin and other antibiotics, and wrote up their work in a number of additional papers. Hazen was also teaching medical mycology to students from city hospitals and others who came to the branch for mycological training, making good use of her collection of cultures and slides. In still another function, she visited local laboratories around the state under the auspices of the New York State Association of Public Health Laboratories, giving practical demonstrations of approved laboratory practices and techniques.

In her off-hours she maintained her long association with Columbia, commuting between the branch and the university by subway and bus until 1954, when the word came that the New York City branch of the Division would be closed and the staff transferred to Albany. In the Division's annual report that year she wrote that during the transition the work was greatly handicapped, but with "no serious break in the continuity." She was more open with her friends, saying, "It set my work back six months."

She was welcomed at the Division in Albany, where she showed keen interest in the projects of her colleagues and offered useful counsel to her juniors. Still carrying on the diagnostic work in mycology formerly done at the branch, she accumulated more material for her book, *Laboratory Identification of Pathogenic Fungi Simplified*, which was first published the following year and dedicated to her mentor at Columbia, Rhoda Benham. Co-author of the book was Frank Curtis Reed, a colleague at the laboratory, who was its illustrator and photographer.

Despite the move, she refused to be separated from either Columbia or her Claremont Avenue address, and took a second apartment in Albany. Now commuting weekly, instead of daily, and by train to the station at 125th Street, she hurried back to New York every Thursday night to her apartment and her work at Columbia, returning to Albany on Monday morning to resume her job at the Division. At the university she had the title of guest investigator, and although the job was unpaid, she felt an obligation for the laboratory space and other facilities made available to her. She was determined to "earn her keep," as she put it.

At Columbia, Hazen carried on her research with *Microsporum audouinii* and, later, a more dangerous pathogen, *Sporothrix schenckii*. She also helped in the diagnostic services in clinical mycology and taught. How well she "earned her keep" was described by Margarita Silva-Hutner, then director of the laboratories, at the Hazen memorial meeting in 1975: "She contributed scholarly lectures to our mycology courses, carefully written and documented from the most recent literature. . . . She also provided benchside instruction in and out of the classroom, giving friendly or stern advice as the occasion required, but always imparting invaluable details of technique, and providing standards of ethics for the young."

Brown, in Albany, kept on with her work on nystatin, further purifying and refining it, finally producing it in crystalline form. She and Hazen did additional investigations of *Streptomyces* microorganisms, discovering two more antibiotics, neither of which proved to be useful in humans. Their joint publications continued to appear in the journals for a number of years. Another of Brown's activities was passing on to Squibb the results of her further work on nystatin as the firm was developing its large-scale production techniques.

The training program at the Division was still another of Brown's assignments, and she was its representative on the

State Health Department's training committee, which was responsible for instruction of professional staff members for the whole department. She arranged summer training programs for college students, brought in outside speakers and set up laboratory seminars for Division professionals. As a recruiter, she visited colleges and high schools to interview promising candidates for jobs, and advised community colleges on curricula that would point students toward later positions at the Division. High school and college groups and service clubs in the Albany area were audiences for her popular talks on science (she was on the speakers' list of the local section of the American Chemical Society), her favorite being "War among the Microbes," which she used as an opportunity to speak of her specialty—antibiotics—stressing the dangers of unnecessary use.

First of the major honors to come to Hazen and Brown was the Squibb Award in Chemotherapy, which was presented in 1955 for "conspicuous accomplishment in the advancement of chemotherapy through their discovery and evaluation of nystatin." The $5,000 honorarium would prove to be the only cash award they received for their work. In accepting the honor for Hazen and herself, Brown spoke of their work as just one more research project carried out at the Division which had come to fruition, adding, "The fundamental knowledge gained by painstaking investigation in a somewhat ivory tower has opened the path to an extended horizon." The photo reproduced on page 86, which was taken after the award ceremony in Albany, is one of the few that shows Dalldorf with "the girls," as he sometimes called them (Hazen was then seventy and Brown fifty-six).

Brown was elected a Fellow of the New York Academy of Sciences in 1957, and the same year Averell Harriman, then governor of New York, sent a congratulatory letter to her and Hazen on the work on nystatin: "You are one of the reasons that the New York State Health Department is a leader in public health in the world." Two years later Brown was the subject of a biographical sketch in *Women of Modern Science* by Edna Yost, first published in English and later in several foreign language versions.[1]

Hazen retired from the Division in 1958, but remained in Albany with an honorary appointment as associate professor of microbiology at Albany Medical College, meanwhile working on revisions to the second edition of her book and maintaining her schedule at Columbia. In 1960 she moved back to Claremont Avenue in her beloved New York, increasing her attention to Columbia to full time, still unpaid, and with the added title of emeritus research professor. Her ties with the Division were not entirely cut, for she still acted as a consultant on mycology.

Hazen and Brown had continued their research with *S. noursei* and other species of *Streptomyces* largely out of scientific curiosity, not in the hope of finding other antibiotics. However, they did discover two more, phalamycin in 1953 and capacidin in 1959. Phalamycin proved to have antibacterial action—of particular interest because it was produced by a variant of *S. noursei,* which had yielded the antifungal nystatin. Capacidin, an antifungal agent but not suitable for human use, was of research interest to other scientists, and in 1959 Hazen and Brown were invited to present their findings at an international symposium on antibiotics in Czechoslovakia. During the visit they were honored guests on tours of pharmaceutical plants in Prague, where they discovered to their amazement that nystatin was being manufactured in huge quantities. They also learned that it was being produced in Russia, in both countries without payment of royalties, for patent protection had not been possible there.

Despite the fact that the fund established with the nystatin royalties was making a number of travel grants for other scientists to attend similar international meetings, both Brown and Hazen traveled to the symposium on their own funds, so adamant were they that they should not benefit personally from their invention.

In 1968, the year Brown retired from the Division (in forty-two years she had taken only two "sick days"), she received the Distinguished Service Award of the New York State Department of Health "for outstanding achievement in the advancement of health for the people of New York State," and a year later both Hazen and Brown were given honorary Doctor of Science degrees from Hobart and William Smith Colleges.

Hazen had been called upon some years earlier to write, with Margarita Silva (later Silva-Hutner), the biography of Rhoda Benham, who had been Silva's as well as Hazen's teacher at Columbia. The biography, which appeared in *Mycologia* a few months after Benham's death in 1957, was at once a saddening assignment for the authors and an opportunity to pay tribute to their friend and inspiration. There was particular significance for Hazen then when in 1972 at a ceremony at the University of Pennsylvania she received jointly with Brown the Rhoda Benham Award of the Medical Mycology Society of the Americas, "for meritorious contributions to medical mycology."

Brown had been recognized by her alma mater in 1957 when she was made an honorary member of Phi Beta Kappa. In 1972, shortly after the Rhoda Benham Award, she was given further recognition by Mount Holyoke—the honorary degree of Doctor of Science. In the citation, her major accomplishments were reviewed, and she was saluted for "your continuing interest in education, and for your dedication to the professional achievement of women."

The crowning achievement for Hazen and Brown came in 1975 at a meeting of the American Institute of Chemists in Boston when they were presented the Chemical Pioneer Award. The presentation was most unusual, for not only did the Institute change its bylaws in order to recognize Hazen, who was not a chemist, but it was the first time the award had been given to women. Hazen was too ill to attend, so Brown accepted for both.

While she dedicated her professional life to science, Hazen kept her personal life to herself, always managing to maintain close contacts with her relatives and with friends she had made in the South and later during her New York years. Her relatives and nonprofessional friends knew quite a different person from the one her colleagues did, and they followed her career closely, taking pride in each new accomplishment and relishing her visits, phone calls, and letters. They learned of her scientific achievements and honors from others, however, for she would not talk of them or would minimize them when asked.

Standing only a little over five feet tall, Elizabeth (as she was known in her New York years) was slight and feminine, yet had tremendous vitality. She always wore makeup, shopped at expensive stores, and, perhaps as a remnant of her college certification in dressmaking, wore beautiful clothes—size six or eight. She went regularly to beauty shops, where she had her hair touched up to preserve its natural tinge of red. Later in life, when her hearing suffered, she bought a hearing aid, which she then refused to wear.

While Elizabeth loved the subways of her day ("they are so nice and warm and light to read on") she also enjoyed driving in the country with Sarah Humphries, directing her friend over a circuitous route from the city to avoid the ten-cent toll on the Henry Hudson Bridge. She also relied on Sarah and Margarita Silva to take her to and meet her at the airport when she was traveling, and once absent-mindedly handed a toll-taker a $100 bill, instead of a $10. She discovered the mistake but by the time they returned to the booth the recipient had no such recollection.

With savings from the farm rental and her job, Elizabeth invested gingerly in stocks, preferring "blue chips," and was fascinated by the stock market. Her thorough reading of the Sunday *New York Times* focused equally on the business section, politics, science, and the arts (she loved the theater and was an eager playgoer). The parties she held at 3 Claremont Avenue were attended mainly by her associates of the College of Physicians and Surgeons, to whom she was known as a gourmet cook.

Mildred Hearsey, a friend and next-door neighbor on Claremont Avenue, recalled that Elizabeth loved to discuss politics and world affairs, always with a liberal viewpoint, ready to see two sides, but holding firm convictions. "She was a true Southerner," Hearsey said, "one of the few in her era who saw the evils of racial discrimination and was outspoken even in the South where her relatives wondered if she was deserting her Southern heritage."

When she was in the South, visiting relatives and friends, she sought rest and relaxation and would not talk about her work. This compartmentalization of her life was epitomized in a letter she once wrote to Rachel Brown explaining her refusal of an invitation to be a house guest of one of their colleagues during a scientific meeting: "I simply can't work and play at the same time."

A cousin, Beulah Townsley, told of her visits to Memphis, where her "sister-cousin" Clara Hazen and others of the family formerly lived. Lee, as they knew her, always asked them not to have parties for her, preferring just to visit with them and take part in their usual lives. She was addicted to Canasta, and to a game called Horse Race, which she would play all day, with only a break for lunch. She would not play bridge, because it required thought. While playing cards she would have a teaspoonful of bourbon in water, which she made last a whole afternoon.

Beulah, who was considerably younger than Lee, remembered vividly that as a child during one of her cousin's visits, she was frustrated by something and blurted, "I can't!" "Lee jumped all over me," she said, "and then went on to tell me, 'I don't ever want to hear you use that word again! There's no such word in the English language. You can do anything you make up your mind to do. Don't say you can't because you can!' "

Lee Hazen continued to take an interest in her young cousin, correcting the grammar in her letters and insisting that she go to college. ("She thought education was the answer to every problem.") Beulah did go to college, thanks to Lee, who provided tuition, books, and other essentials.

Beulah and Conway Dickey, another cousin, inherited Lee's personal effects, among them her books that Conway gave to Mississippi University for Women. A memento cherished by Beulah is an invitation to the White House issued to Lee for five o'clock on the afternoon of May 13, 1948. A participant at the Fourth International Congress on Tropical Medicine and Malaria held in Washington, Lee was one of the women scientists invited to tea that day by Bess Truman.

Rachel Brown's family, appearance, and extraprofessional activities were very different, her social life and Hazen's touching only irregularly. Brown was taller, perhaps five feet four, more heavily built, and more athletic, and while always carefully groomed, less interested in the latest clothes. Having no outside source of income, she also had to be more frugal. But their friendship was very warm, and in Hazen's later years, Brown's was the strong arm on which she leaned when they were together.

Like Hazen, Brown loved the theater and was a voracious reader. In addition to the scientific literature, she regularly read *The Atlantic Monthly, National Geographic Magazine, Saturday Review* and, in recent years, *Smithsonian* magazine. Unlike Hazen, she loved to play bridge and was an ardent gardener who, in the words of one friend, "delighted her neighborhood with a most attractive array of flowers." In her later years, after suffering a broken leg, she was distressed because she could no longer comfortably kneel to tend her garden.

After their years in college at nearby campuses, Rachel and her brother Sumner were separated geographically, but remained close, as they were in years also (Sumner was just eleven months younger and Rachel liked to say, "For one month a year we're twins"). Their paths separated, when she went west to the University of Chicago and he, after his graduation from Amherst, went farther east to start his ecclesiastical training at Episcopal Theological Seminary in Cambridge, Massachusetts. Then he went far west, taking up his ministry as a missionary in Prairie City, Oregon, and later came back to serve churches on Cape Cod, moving in 1955 to Wilbraham, a village near their old home in Springfield. There, at Wilbraham Academy (now Wilbraham & Monson), he was chaplain and instructor in religion and mathematics until his retirement in 1971. The move brought him and his wife Ruth within a few hours' drive of Rachel's home in Albany, and there was frequent visiting back and forth.

The dedication to their church shown by Sumner and earlier by their mother—who supported the family for years by her work in religious education—was also a part of Rachel. She joined and became active in St. Peter's Episcopal Church in Albany shortly after moving there in 1926, and it was there that her friendship with Dorothy Wakerley developed. Her other main connection with St. Peter's was its Sunday school, in which she taught for over forty years, giving love as well as instruction to the children and, as they grew up, to their children.

Those who watched her at the church school and who knew her at home with visitors and the youngsters in the neighborhood marveled at her way with children. Dorothy Wakerley recalled a day when Rachel was being visited by a little boy and girl as she was working in her garden. The little girl kept trying to get the boy to come and play with her, but he resisted, being more intent on helping Rachel. Finally in exasperation at his friend's insistence, he said, "I don't want to! I want to play with this girl."

Rachel never forgot the beneficence that allowed her to go to college, or her mother's admonition: "Never buy anything you can't pay for." By living carefully, avoiding charge accounts and credit cards, and religiously adding to her savings, she was able to "repay Aunt Etta" vicariously—and very quietly. When pressed, she once admitted grudgingly that she had helped several young people through college but refused to give details. For over twenty years, through the Christian Children's Fund, she helped support one or two children a year in Southeast Asia and India. And, in addition to the time and effort expended on church affairs, she contributed generously.

Other beneficiaries of her open-heartedness were members of her nonrelated family, which was as large and more diverse than her own. She called it her "Chinese connection," and it began in 1946 when one of Rachel's friends was seeking lodging for a woman physician from China who was a visiting scientist at the Division. Rachel's mother was still living with her and Dorothy, but her grandmother had died earlier that year, leaving one room vacant. Jessie Hendry, a classmate of Rachel's at Mount Holyoke and a fellow scientist at the Division, asked whether the room could be made available to the visitor from overseas. Annie Brown and Dorothy were as pleased as Rachel to extend their hospitality, and Lin Fei-ching, who was studying under Hendry, joined the household until completion of her visit to the Division six months later.

When Fei-ching's room was vacated, Rachel was asked if it might be opened to another visitor from China to the Albany laboratories. Again the Browns and Dorothy were happy to welcome a new member to their home—Chen Sing-mei. Sing-mei had expected to stay only until her mission at the Division was completed, but her residence was extended after she had finished her work there and joined the Microbiology Department of Albany Medical College. Annie Brown died in 1958 and the other three continued at 26 Buckingham Drive for another twenty-two years.

Like Dorothy and Rachel, Sing-mei was unmarried, but after her arrival the home had no longer lacked children. One of Sing-mei's brothers, his wife, and infant son came from mainland China to join them, the wife and child staying for some months. They were followed by four of Sing-mei's nieces and a nephew, who came to the United States over a period of years to go to college. As they arrived for short or longer stays, they were gladly adopted by Rachel and Dorothy, and given a new home base to which they returned during vacations and holidays. The house itself was gradually equipped for children as the nieces and nephew—and later their sons and daughters—came home to visit. Rachel's own nieces and their parents were equally welcome visitors, making for some interesting problems of scheduling and logistics in a four-bedroom house. But these problems were pleasures for Dorothy and Rachel, adding joy and excitement as various members of the large family arrived at their "home away from home," as they called it.

The third member of the Brown-Hazen-Dalldorf team left Albany in 1957 when he resigned as director of the Division of Laboratories and Research and took on a brief assignment as research director of the National Foundation (March of Dimes). Anna Sexton, in her chronicle of the Division's first fifty years, wrote of the two developments for which Dalldorf was best known during his twelve-year tenure: "There is seldom an opportunity such as was had in the decade from 1947 to follow daily, intimately, the discovery, developing significance, and universal recognition of a new disease entity, as in the Coxsackie virus infection, and a new therapeutic product, as in nystatin."

Dalldorf's accomplishments were widely recognized outside the Division as well. He had received an honorary Doctor of Science degree from Bowdoin College in 1953, was elected to the National Academy of Sciences in 1955, and in the same year received the Fisher Memorial Award of the American Chemical Society. The New York University College of Medicine gave him its Distinguished Service Award for 1956 for his work in infectious diseases and pathology and, in particular, for identification of the Coxsackie virus. In 1957 the German institution at which he had held the fellowship in pathology thirty years earlier, Albert-Ludwigs-Universität, awarded him an honorary doctorate at its five hundredth anniversary—an honor he considered to be his greatest distinction.

In 1959, after he became a member of the Walker Laboratories of the Sloan-Kettering Institute for Cancer Research, of which he was also a trustee, his work on the Coxsackie viruses was further rewarded. He received that year the Albert Lasker Award in Medical Research from the American Public Health Association, and in 1964 was awarded the Medal of the New York Academy of Medicine.

Dalldorf resigned his active role at the Walker Laboratories in 1966, but continued as an emeritus member, carrying on epidemiological studies of Burkitt's tumor which he had started earlier in Africa, to which expeditions Frances Dalldorf also contributed her services. These studies, which were supported by Sloan-Kettering and in part by the Brown-Hazen program, were in pursuit of the hypothetical "cancer virus," but his investigations did not tie down the suspected linkage between a form of cancer and mosquito-borne virus diseases. He continued to believe, however, that there might be some such link.

The Dalldorfs' life became seminomadic after he left the Division, the *Wanigan,* a thirty-foot motor sailer, serving as their sometimes mobile home, being berthed in Florida or the Chesapeake Bay area while they were on their scientific expeditions. Sailing had supplanted Gilbert's earlier avocation of flying, which had started in their Westchester County days. There he learned to fly and bought his first plane, a 65-horsepower Luscombe, which was succeeded by a four-place Stinson when they lived in Albany. After 1956, sailing the waters of the east took the place of cross-country plane trips, and they found the eastern shore of Maryland so much to their liking that in 1963 they decided to establish a land base there. In Oxford, which had been a major port before Baltimore but had later languished, they found a waterman's house across the road from a boat yard and a dock for *Wanigan.* Three years later the useful days of that craft were ended by an accident on the intercoastal waterway in Virginia. *Wanigan II* then took her place at the dock, in races of the Oxford Sailing Association, and on cruises up and down the Atlantic coast. Ashore the Dalldorfs pursued active retirement with their medical, scientific, and gardening interests when not visiting or being visited by family and friends.

8. *Evolution in Philanthropy*

After early 1957, when the Brown-Hazen program of grants was formally inaugurated, Hazen, Brown, and Dalldorf had a new and consuming interest. With a modest amount in nystatin royalties having already been received for the program and the outlook good for continuation, Research Corporation officers and the three founders agreed that the time had come to start the grants program.

From the start, the program required a reconciliation of differing viewpoints in merging the interests of Brown, Hazen, and Dalldorf with the philanthropic and granting experience of the foundation as the administrator of the program. The founders were acknowledged authorities in their respective fields, determined to apply the program's share of the nystatin royalties to the advancement of biological science. The officers and grants staff of the foundation knew the problems and pitfalls of program management, and were familiar with the techniques of assembling and interpreting the mass of information on which informed and objective grants judgments could be made.

It was to be expected that there would be differences between the founders and the foundation staff as to how best to bring into reality the objectives of the founders. Dalldorf and the Research Corporation representatives were frequently at odds on both substantive and procedural matters, often resulting from his penchant for informal actions and the foundation's need for at least minimal documentation. Much to the credit of the committee set up to advise on the program, it was seldom that these differences were not reconciled to the satisfaction of all.

The members of the Brown-Hazen Committee as the program went into operation in 1957 were Brown, Hazen, and Dalldorf; Herman E. Hilleboe, commissioner of health of the State of New York; Sam C. Smith, associate director of grants of Research Corporation and the foundation representative; and Currier McEwen, professor of medicine and former dean of the School of Medicine of New York University, the "public" member. All were mandated by name or by function in the agreement between the donors and Research Corporation, except for the public member, chosen by the others. Dalldorf was elected chairman when Hazen declined the post, and Brown was secretary.

McEwen, whom Dalldorf had first known at Bellevue Hospital when they were medical students, proved to be an extraordinarily fortunate choice. He brought to the committee a broad view of medicine, science, and education, as well as long experience as a physician and researcher, as a teacher and manager, and in the grants area as a grantee, a grantor, and a grant administrator. Charles H. Schauer, vice president of Research Corporation and its chief grants officer, served as the foundation member only at the organization meeting of the committee in April 1957, yet his continuing counsel had a pervasive influence on the program during its twenty-one years of operation.

Hazen, Brown, and Dalldorf constituted the steering committee, which carried the authority of the full committee to make grant recommendations in certain special situations. Each of the three spent countless hours poring over the grant applications as they began to come in, reviewing the pertinent scientific literature, consulting with colleagues and, on occasion, visiting potential grantees to gain fuller information and understanding of the work proposed. As the program got into full swing, they met two or three times a year for one- or two-day meetings of the full committee, at which applications (often fifty or more) were taken up one by one, discussed, and voted on. It was arduous and time-consuming but to the three founders a labor of love.

At the outset of the grants program the disciplines to be supported were announced as biochemistry, microbiology, and immunology. Priority would be given to certain special needs of the Division of Laboratories and Research, with which all three founders had been associated and at which the work on nystatin was done. Medical mycology—naturally of prime interest to Brown and Hazen—was not singled out for particular attention, since it was covered under the three broader categories and the committee did not want to restrict too tightly the fields of research it hoped to advance.

The kind of work to be supported was basic research: developing through intuition tested by experiment the bits of new scientific knowledge that would supply some missing links or provide fresh insights into biological processes. Few direct breakthroughs like nystatin were expected from this support, but it was hoped that when added to the storehouse of scientific data available to researchers everywhere, these contributions might lead in time to some practical measures for combating human disease.

The investigators to be supported were faculty and staff members in colleges, universities, and medical institutes, teacher-researchers with original ideas, with unfettered imaginations, and with full freedom to follow their own independently conceived research interests. The academic environment of institutions at which the proposed work would be done would have an important bearing also, since this kind of creative effort could take place only where there was an atmosphere that not merely permitted but encouraged such independence.

The Brown-Hazen fields of interest extended Research Corporation's granting activities into the medical-biological sciences, but the Brown-Hazen guidelines paralleled to a great extent those that had been worked out over the years for the foundation's grants programs in the physical sciences. They provided, and deliberately so, a number of intangible factors for the advisory committee to weigh when making grants judgments—factors that might influence as much as the more measurable ones the chances for research projects to yield significant results.

Sam Smith, who as Research Corporation member of the Brown-Hazen Committee helped establish the program, reflected recently on one of the problems to be faced in setting up a new grants program. He told me: "It's a cliché among foundation people but nevertheless still true that giving away money wisely is a very hard thing to do. You can invest in experienced individuals who have proven track records and expect to get results of high order. Or you can bet on promising youngsters, helping them to establish themselves in their chosen fields and bringing along a new generation of seasoned performers. The former is a low-risk investment with almost certain payoffs; the latter is high-risk but eminently worthwhile for those cases where the youngsters develop into true professionals making real contributions in their fields."

Commenting on grants for scientific research, Smith added: "If you're looking for big or almost certain advances, you pick the people with established records. But, on the other side, if you don't pick some of the upcoming generation of scientists, they won't ever have the chance to prove themselves. Research provides this chance—the opportunity of actually *doing* science, not just reading textbooks, but practicing the art. There are many frustrations in research and they will grind down all but the most dedicated. Tested in this crucible, young people with fresh minds and few inhibitions come out much better prepared to find answers to some of the major unsolved problems of science."

As it developed, the Brown-Hazen program supported both fledgling and seasoned researchers. In the beginning it sought out scientists who were known to committee members or their colleagues for work already accomplished in biochemistry, microbiology, and immunology. Representative of the kind of potential grantee being sought was Max Delbrück of the California Institute of Technology, who received one of the earliest Brown-Hazen grants in 1958 for research which later proved to be instrumental in advancing molecular genetics. Delbrück was a well-established investigator at the time of the grant and had, in fact, already done the work for which he would receive the Nobel Prize in Medicine eleven years later.

Such bright stars were not very numerous, however, and many of them were already well supported by federal funds. Yet the younger investigators seldom were able to get such funds for their research. Also, since they were less well known, they were harder to find. To aid in identifying them, the Research Corporation regional directors, whose normal function was making similar searches for prospective recipients of grants in the physical sciences, were enlisted to seek promising candidates for Brown-Hazen grants. In their travels to institutions of higher education across the country, these field representatives now added as places to be visited medical schools, as well as departments of biology in colleges and universities. In these informal visits, generally made without advance notice, they sampled the research atmosphere of the institutions and departments and talked with senior faculty members about new appointees to the staff who had research potential but lacked the funds to start their own work. On finding investigators whose projects they deemed appropriate and sufficiently promising, they encouraged grant applications.

The information they gathered and passed on to the Brown-Hazen Committee—either in reports or at committee meetings—provided firsthand, current, and often subtle insights into potential grant situations. To these were added the more specific scientific data furnished by the applicants themselves on the purpose, procedures, and significance of the work proposed. Another source was judgments sought from qualified outside referees on the importance and feasibility of research in their fields of specialization. With all this data at hand, the committee had abundant information for its recommendations to grant or deny individual proposals.

The committee had decided at the beginning of the program to adjust the level of its granting expenditures to the actual income received by the Brown-Hazen Fund from the nystatin invention. Some $75,000 was available for grants when the program got under way, and, as nystatin began to find wider usefulness as an antifungal drug, the royalties increased slowly, by 1960 running about $100,000 a year for the Fund. Plans were then made for a slight extension of the program into a new area—grants to biology departments to encourage research by undergraduates.

Research Corporation had long maintained an interest in the liberal arts colleges as producers of well-qualified students going on to take advanced work in science, many of the country's outstanding scientists having taken their undergraduate work in these four-year institutions. Results of an experimental program in the physical sciences had indicated that upgrading whole science departments in such institutions could increase their production of broadly trained graduates prepared and motivated to go on to careers in science. It was felt that grants for biology departments from the Brown-Hazen program could have the same effect for that discipline.

The first Brown-Hazen grant of this kind was made in June 1960 to Hamilton College in Clinton, New York, "to stimulate an interest in independent research in biology at the undergraduate level." Almost before it got started, however, this effort—and indeed the whole Brown-Hazen grants program—was threatened by a development in the pharmaceutical industry.

The patenting and licensing specialists at Research Corporation—the group that had worked out the arrangements with Squibb for the development and sale of nystatin—had estimated at the beginning of the Brown-Hazen program that total royalties from the antibiotic might amount to $2 million before the patent expired in 1974. In making the estimate they were well aware of the danger of such forecasts, knowing that there was always the possibility of another, newer drug coming to the market to displace an older one. The 1960 threat to the Brown-Hazen program was an announcement by Squibb that it was introducing a new antifungal drug called Fungizone, in what seemed to Research Corporation to be direct competition with nystatin.

Smith alerted the other members of the committee to the threat and, in a memo to foundation officers, invited them to a meeting of the committee to be held on October 21, 1960, at which "a wake will be held to mourn the passing of the Brown-Hazen Fund." He added that there would have to be a drastic curtailment of activities until the income picture became clearer, but that if a bare minimum of new obligations was taken on, "the entire program can probably be brought to an orderly close with fairness to all."

At the October meeting the committee heeded the warning and recommended only minimal grants expenditures, as it did again at its next meeting in mid-1961. Then it became evident that the threat would not materialize, that the nystatin royalties would not be severely affected, and that additional Brown-Hazen grants could be made.

Fungizone, the Squibb name for amphotericin B, proved to have the superior qualities claimed, in that it was more soluble than nystatin and could be injected into the bloodstream to be carried to the parts of the body affected by the deep fungus diseases which nystatin could not reach. Nystatin, however, continued to have its own special uses, a major one being to combat fungus infection of the intestinal tract after treatment with antibacterial antibiotics. There was only a slight drop in royalties in 1961, then the returns from the nystatin patent continued to rise until the expiration of the patent in 1974.

The Fungizone cloud having passed, the committee was able

to consider the applications that had been generated for project grants, as well as a few proposals for the aborted departmental grants program. Although this latter phase of the Brown-Hazen program resumed after 1961, it never became one of the program's major activities. Between 1960 and 1969 there were ten grants, totaling about $355,000, made for upgrading biology departments in the liberal arts colleges, ranging from some $6,000 for the grant to Hamilton College in 1960 to $85,000 for Colorado College in 1967. Most were given independently of the Research Corporation program in the physical sciences, but three went to colleges to which the foundation had already given grants for chemistry and physics: Lebanon Valley College in Pennsylvania, Seattle Pacific College and Pacific Lutheran University, both in Washington.

The $35,000 Brown-Hazen grant to Pacific Lutheran University in 1968 was a particularly interesting one in that it helped fund a cooperative research and teaching program of two academic departments which had recognized the kinship of modern-day biology with chemistry. The program marked the introduction of molecular and cellular biology at an institution which previously had, like many others, offered only the classical type of training in biology.

A Research Corporation departmental grant made two years earlier for the natural sciences division had helped build a firmer base for the sciences as a whole and had strengthened both the biology and chemistry departments by the addition of new and enthusiastic faculty. Two of the newly recruited teacher-researchers, JoAnn S. Jensen, a physiologist, and Burton L. Nesset, a biochemist, proposed for a Brown-Hazen grant a further program having as its primary intent the stimulation of student participation in independent study and research projects; the program also was expected to give fresh impetus to the faculty to do research in cellular biology.

The grant proposal noted that while it would be inappropriate for an institution of the size of Pacific Lutheran, then with an enrollment of about 2,200, to undertake an extensive program in these areas, all students of chemistry and biology should have the opportunity for at least a moderate exposure to cell study with its attendant instrumentation and techniques. "The staff of both departments believe that a student can develop competence in the sciences most readily by trying his own hand at experimental work," Jensen and Nesset continued in their proposal. "Thus, a sort of guiding principle in the instructional program is to give students maximum laboratory experience consistent with availability of funds and personnel."

Reporting on the program in 1978, ten years after its initiation, Jensen noted that it would be difficult to separate completely the effects of the Brown-Hazen award from those of the 1966 Research Corporation grant, because the earlier grant had set the stage by providing additional personnel. "Growth produced more growth," she wrote. "With the nucleus of faculty and equipment obtained earlier, we have been able to attract more good faculty and obtain more sophisticated equipment." She added that where ten years before few of the chemistry and biology faculty were engaged in research, most were currently involved, together with their students.

The Brown-Hazen funds, plus those the institution pledged as partial matching, made possible the establishment of a number of new courses which provided fresh experimental possibilities for the students. The students responded quickly to the new offerings, and some went still further to elect independent research projects, with both publications and presentations to regional professional societies resulting. "An expected spin-off from all of this," Jensen reported, "was a noticeable increase in the number of students entering graduate school in this general subject area."

Departmental-type grants from the Brown-Hazen program were also made to Albany Medical College in 1963 and 1966 for strengthening and expanding the biological sciences. Including an earlier grant for the department of biochemistry, a total of

$148,600 was given to the institution to broaden its research and teaching capabilities. The grants were equally beneficial to the college and to its neighbor, the Division of Laboratories and Research, with which it had had a long and mutually helpful relationship, students and staff of both institutions being able to transfer freely between the two to take advantage of the specialties offered at each.

Toward the end of the 1960s, Research Corporation began phasing out its program of departmental grants, after other agencies, both federal and private, had started larger but similar programs aimed at essentially the same group of institutions. The Brown-Hazen Committee willingly followed the foundation's lead in the phase-out, deciding in 1968 to accept for further consideration only those proposals prepared by colleges previously invited to apply for the grants.

Members of the committee had not been uniformly enthusiastic about these grants; there was some feeling that the Brown-Hazen funds should be used for research in biology with more direct medical application. From the Research Corporation viewpoint, the Brown-Hazen departmental grants, like that program's project grants, had allowed the foundation to enlarge its fields of scientific interest to include a major area of modern science, an influence that continued after the last of the Brown-Hazen grants had been made. Today biology proposals, as well as those in the physical sciences, are considered in one of the foundation's regular grants programs.

Another phase in the evolution of the Brown-Hazen program had been initiated in 1966, also as a result of the Research Corporation affiliation. Hal H. Ramsey, foundation grants representative for the western states, also had in his territory both Canadian and Mexican institutions, whose applications were considered on the same basis as those from domestic colleges and universities.

On his visits to Mexico he had found several academic centers where the greatest ferment and enthusiasm for research was found among the young biologists and biochemists. Most of them had taken their graduate work in the United States and, refusing to join the "brain drain" movement, had returned to Mexican universities to apply their talents for the advancement of their own country. While anxious to start research in their chosen areas of science, these investigators could get only minimal support from their institutions and none from federal agencies; their only chance to do research was in assisting in projects of interest to the senior people to whom most funding was directed. Their situation was even more severe than that of the younger investigators in U.S. institutions, but since they were in the biological sciences, they were not eligible for support under the Research Corporation physical sciences grants program.

After the entry of Research Corporation into the medical-biological sciences via the Brown-Hazen program, Ramsey encouraged the fifteen to twenty Mexican scientists whom he deemed most promising to prepare applications for grants. Nine of this group received grants between 1966 and 1969, competing successfully with researchers from U.S. colleges and universities on the merits of the science proposed. A total of $65,370 of Brown-Hazen funds was expended for their projects.

Reporting in 1970 after the conclusion of the Mexican mini-program, Ramsey cited the accomplishments at the Instituto Politecnico Nacional and at the Universidad Nacional Autonoma de Mexico, the two institutions to which the grants had gone. One whole department of immunology had been created around a grant recipient and two part-time students, and currently had twenty full-time people doing research. A microbiology section had grown from one investigator doing no research to a group of twelve, all actively engaged. More than twenty papers on Brown-Hazen-supported work had been published in reputable scientific journals, over fifty students had been trained, and most of the grant recipients had later received funds for continuation of their work from other sources.

Among the sources of the additional funding for their research were the two institutions themselves, a group of local businessmen and two agencies of the Mexican government, all presumably encouraged by the example of the Brown-Hazen grants. The Consejo Nacional de Ciencia y Tecnologia, a newly created federal agency, modeled to some extent after the U.S. National Science Foundation, was headed by a scientist who earlier had been awarded a Brown-Hazen grant. Possibly reflecting this influence, the agency had broken through the European–Latin American tradition—the pyramid system of awarding funds to senior professors with some hope that there would be a dribble down to the juniors—and was making grants on a competitive basis to younger, less-established scientists for their own independent research.

Summing up with enthusiasm he still thinks is justified today, Ramsey wrote in a 1970 report to the Brown-Hazen Committee: "This effort in Mexico is, to me, the quintessence of what a foundation should do. . . . The committee agreed to give it a try even though the risk of failure seemed much higher than in supporting people and places which were intimately familiar. . . . You [the committee] have assisted intelligent, well-trained, industrious people to establish themselves in the scientific world."

Like the departmental phase, the Mexican effort was not an entirely rewarding experience for Brown-Hazen Committee members. The main objection was to the use of nystatin royalties outside the United States when domestic needs were unfilled. However, the contrary view prevailed: the objective was to support good science in the areas of interest of the program, and good science should be supported wherever it was being done, provided that the committee had full information upon which to act. Ramsey had seen to it that this kind of information was available.

A long-term personal interest of both Hazen and Brown was the urge to do something to encourage women in the sciences. Brown in particular was frustrated in finding that not many college women had the same kind of interest in science that she had had at their age. After studying the record of the Brown-Hazen program up to 1974, she commented sadly in a letter to Sam Smith: "Its funding for scientific research almost exclusively goes to men. Is it so hard to find worthy women?" At about the same time she wrote to Hazen, "I am disappointed that women still are not pushing themselves in science." It was a thought that continued to trouble her.

Some five years earlier, attempting to find some answers, she and James S. Coles, president of Research Corporation, had made a series of visits to eight women's colleges to see what kind of aid the Brown-Hazen program could give either to whet an appetite for science among women, or to help those already interested to go on to advanced study. The results were discouraging, for, as Brown said later, "We went around to the campuses, but couldn't stir up much interest. The trouble seemed to be lack of initiative on the part of the women."

There were some positive results from the brief campaign, however. One application that was funded was from Mount Holyoke for a contribution toward purchase of an electron microscope and the recruiting of a specialist to operate it, with the thought of using the sophisticated instrument as a teaching as well as a research tool. The grant accomplished its purpose, for there was an upsurge of undergraduate interest in biology, which continues today.

Another project aimed toward advancing women in the sciences was a summer program of undergraduate research at Vassar College in 1969 and 1970. This project brought together three professors and six undergraduates to work on research programs of the principal investigators. The object was to provide practical education in biology in the crucial early years of college and to stimulate talented young women to seek careers in research. At the end of the first summer, two first-year students elected independent research for academic credit in their sophomore year, and three continued to work on the faculty research projects. Only one of the six had been committed to research at the beginning of the summer session, and after their laboratory exposure there were five.

With these and two other exceptions, women applicants for research funding were given the same treatment as men, as was the case in Research Corporation's other grants programs. If the project proposed appeared to have scientific significance, if the referee comments were favorable, and if the advisory committee saw promise in the work, funds were granted without regard to gender. (It was Brown's belief, however, that in general the applications from women were "very well prepared, even superior!")

The other deviation from this general policy came when the final awards from the Brown-Hazen program were being made in 1977. There were two that were strictly sexist—the funds for scholarships and fellowships given to Mount Holyoke, named for Brown, and to Mississippi University for Women, in Hazen's name—both restricted to advancing the studies of women in science.

9. Concentration in Mycology

In its several stages of evolution before 1970, as described in Chapter 8, the Brown-Hazen program had held pretty much to its announced fields of grants interest—the biological sciences in general, and microbiology, immunology, and biochemistry in particular. Medical mycology had not been singled out for special attention, but proposals for research on fungi and the fungus diseases had always been encouraged and, when they came up to program standards, funded. Still, Hazen and other members of the committee were puzzled at the relatively small number of grant applications coming in for work in medical mycology. It was especially hard to understand since there had been a decline in even the very limited federal support that had been available for such work.

By the end of the 1960s royalties for the Brown-Hazen program had increased to over $500,000 a year, but the committee looked ahead to the mid-1970s when the patent would expire and all income cease. The next phase of the program would probably be its last, and since the committee sensed an unfilled need for heavier support of medical mycology, it seemed fitting that the remaining resources of the Fund be brought to bear upon the problem that had first engaged the two women scientists more than twenty years earlier. An informal study was undertaken to assess the current needs of medical mycology with a view toward applying Brown-Hazen funds to those needs.

The study led to two general conclusions: the Brown-Hazen program should reassert the public health importance of the fungus diseases to help attract other funds to support the largely neglected field and more people to work in it; the program itself should place new emphasis on training to increase the numbers and the competence of those working in medical mycology. The means to accomplish this and the specifics of the new program, however, were matters of some controversy.

The membership of the committee had changed little since its formation in 1957. Hollis S. Ingraham had succeeded Hilleboe as New York State Commissioner of Health in 1963 and, by terms of the original agreement, had also taken his place on the Brown-Hazen Committee. Ramsey relieved Smith as the Research Corporation member from 1967 to 1969, after which Smith took up the duties again until 1971, when he was succeeded by Kendall W. King, the foundation's assistant vice president for grants. Dalldorf as chairman, Brown as secretary, Hazen, and McEwen continued throughout as active members, rarely missing a meeting.

The medical mycology program got under way in 1970, and initial grants were made for training as well as research, but there continued within the committee the old debate as to the kinds of research to be supported. A main subject in these discussions was basic versus applied—whether the emphasis should be on fundamental work, with the payoffs, if any, coming sometime in the future, or for work that had more immediate application.

Currier McEwen, an ardent partisan of basic research, recalled later the view he expounded during these sessions: "If you never do anything but apply knowledge you already have, pretty soon you'll run that vein out. You can only make progress in the future by turning up new basic information."

Elizabeth Hazen had a different view. In a rare resort to formality, she delivered to the committee in June 1972 a written statement reflecting her impatience with progress in her field. Pleading that the immediate need in medical mycology was for early diagnosis, specific treatment, and prevention, she wrote: "We have some of the finest and best-equipped mycologic laboratories in the world and we have highly trained and gifted mycologists. . . . But what we have not done is to improve the treatment of the infected person. That is the crying need in mycology."

McEwen, in his customary role of statesman, composed the differences in a resolution he offered at the next meeting of the committee: "The program policy of the Brown-Hazen Fund beyond 1973 will be restricted to support of mycology, stressing research, training in research, and medical applications." The resolution was adopted unanimously.

The new policy accomplished several things. By narrowing the fields of interest to combat of the fungus diseases, it concentrated the funds on a smaller area where they could have greater impact. Within this limited field, however, it encompassed basic research programs to develop new knowledge, training programs to prepare other researchers for like work in the future, and practical programs to apply existing knowledge of the fungus diseases toward the alleviation of human suffering. All of these were parts of the Brown-Hazen effort as it went into the last stage of its grants program.

During the study that led to the new direction for the program, talks with some of the leading spokesmen for medical mycology had revealed, among other problems, that the field was not attracting young people. Reasons given were the meager funding available for mycological research, the few centers equipped and staffed to offer sound training, and the rather dismal outlook for trained people when they started looking for jobs. Believing that this cycle had to be interrupted at some point, the Brown-Hazen Committee decided to help strengthen existing centers by making grants for research and training, believing that there would be jobs once the people were trained.

High on the list of the kinds of skilled people needed were physicians who would look for and diagnose the often-unsuspected fungus diseases, and laboratory technicians who could test for and identify disease-causing fungi in the specimens sent them for analysis. Also, biological-medical researchers needed to be alerted to the importance of fungus diseases as a challenging field for study and given experience in the investigative techniques that could lead to better understanding of pathogenic fungi, conditions under which they could cause diseases in humans, ways to prevent the infections, and means to combat them once they are established.

The first two of these needs were pointed out in a 1971 statement of Howard W. Larsh of the University of Oklahoma, one of the relatively small group of scientist-academicians attuned to the potential threat of the fungus diseases. He asked: "Where do physicians, technicians, and other laboratory personnel receive their training as to significance of fungus infections?" As to the physicians, he noted that in many medical schools the curriculum did not include lectures in medical mycology, in others the microbiology course might have three lectures which cover the entire field of mycology, and in some the microbiology staff considered a guest lecturer for an hour every other year sufficient to orient their students in medical mycology. Even worse, Larsh added: "Most medical technologists and public health microbiologists receive no training in medical mycology. It is not at all uncommon for a graduate registered technician to remark that he was not permitted to examine fungus cultures, as they might be pathogens."[1]

A parallel observation—which still can be made today—was that those who were specializing in mycology were concentrating largely on the nonpathogenic varieties of fungi. While these are of themselves of great scientific interest, they do not cause diseases in humans, and the ones that do—the pathogenic fungi—were the subject of comparatively little research. There is no question that working with pathogenic fungi can be dangerous, for if they can indiscriminately cause disease in the general population, they pose very direct threats to those working with them in the laboratory. Here again training was called for; with understanding of the inherent dangers and with the use of proper equipment and facilities, the dangers of working with pathogenic fungi could be minimized.

Brown's determination to encourage women in the sciences, and her interest in supporting young researchers in a broad range of the biological sciences, made her less enthusiastic than Hazen when the Brown-Hazen Committee decided to restrict the grants program to the combat of fungus diseases. Hazen was strongly—even passionately—concerned about the lack of knowledge in medical mycology and the "deplorable" paucity of training in that specialty and was one of the most avid proponents of the change of direction. Brown went along loyally, however, and soon became one of the committee's most active members in seeking out prospective grantees for the new program.

The committee having decided to emphasize training, the logical first step was to identify the centers and the individuals around the country that were currently involved in or interested in medical mycology. Again called upon to aid in the search were Research Corporation's regional directors of grant programs: Ramsey in the Far West; Brian H. Andreen in the Midwest; David G. Black, Jr., in the Northeast, and Jack W. Powers in the Southeast, who was later succeeded by R. Scott Pyron. Scouring their territories for evidence of ongoing work in this specialization, they were joined by members of the Brown-Hazen Committee when they found likely sites for further exploration. Strangely, they found in some cases that the most promising candidates for grants needed prodding, so unaccustomed were they to any private funding agency's being interested in supporting work in medical mycology. Within a few months, however, applications for training grants began to come in.

As was expected, these applications were for larger sums than were usually requested for the research project proposals the committee had been judging over the years and required support for more than one year. Most project grants had been running from $2,000 to $8,000, almost always on a one-shot basis. The budgets for the training grants ran into tens of thousands of dollars per year for programs that would run several years. But since the nystatin royalties were now coming in at an increasing rate for the Brown-Hazen program, the committee felt that it could not only make larger grants, but for longer periods.

Another consideration was that the training grants, despite their higher costs, would have a greater multiplier effect than would the project grants. Where a grant for a single research project would generally aid in the training, as well as the research, of one principal investigator and perhaps one or two graduate students, a training grant would reach a larger number of teacher-researchers, many of whom would then go out to other institutions to train still others. These grants would also reach workers other than researchers, many of them employed directly at the first line of defense—the physicians, clinicians, and laboratory technicians who could be trained to identify fungus diseases in patients, thus providing the clues needed for proper treatment.

Research and training in research then became the key elements of the Brown-Hazen program as it shifted into high gear for the final phase of its work—a frontal attack on the fungus diseases.

10. Spreading the Word

Several of the programs started with Brown-Hazen grants after 1970 were aimed both at bringing better-trained people into the fight against fungus diseases and at supporting research projects in medical mycology. These programs, which varied considerably as to the emphasis on the research and the training components, were conducted at the College of Physicians and Surgeons of Columbia University, Temple University Health Sciences Center, Harvard School of Public Health, and University of Michigan Medical Center.

Two others, funded initially in 1973, were of particular relevance because they tied in more firmly with the new Brown-Hazen orientation toward training. They also presented some interesting challenges because of the very different kinds of people to be trained. One was at the University of Kentucky in Lexington, a program involving relatively short-term advanced training for a large number of people already working in the field. These trainees, mostly laboratory technologists, were expected to go back to their regular jobs with new expertise to apply to diagnosis of fungus infections, and to give on-the-job training to others at their laboratories. The other was at Washington University in St. Louis, an intensive multidisciplinary postdoctoral program for a much smaller group with M.D. or Ph.D. degrees. Those trained here would be well equipped for faculty positions at other universities or medical centers where they could conduct their own research in medical mycology and bring their specialized knowledge and experience to succeeding classes of trainees.

A large factor in making the first grant to Washington University was the enthusiasm shown by the referees who reviewed the grant proposal, all of whom were authorities in medical mycology. What appealed to most of them was the program's multidisciplinary aspects, which would expose the trainees to medical mycology, immunology, biochemistry, and molecular biology and would link clinical and diagnostic training with opportunities for the postdoctoral group to become actively involved in independent research projects.

Other favorable factors were the caliber and the diverse backgrounds of the medical scientists who would direct the program and the facilities that would be available. The principal investigators were George S. Kobayashi, associate professor of medicine at the Washington University School of Medicine and consultant in mycology at Barnes Hospital, one of the associated units in the university's complex, and Gerald Medoff, associate professor of medicine and chief of the Infectious Disease Division at the medical school. They were joined in direction of the program by David Schlessinger, professor of microbiology at the university. The fields they represented were, respectively, medical mycology, infectious diseases, and molecular biology—a combination to foster the multidisciplinary thrust.

The facilities and equipment available were those of the Divisions of Dermatology and Infectious Diseases and the Department of Microbiology, as well as those of Barnes and three other hospitals in the medical complex, a clinic, an animal research facility, and a biological hazard area for work with pathogenic fungi. The university-hospital linkage would give trainees the benefits of working in an academic atmosphere closely integrated with clinical medicine.

The program provided two years of instruction and experience to equip trainees for academic careers and clinical investigation—an approach that appeared highly promising. While the trainees would be exposed to clinical work, the main thrust of the program was to develop their capabilities for independent research and, through their research with senior investigators, to expand the fund of basic information about pathogenic fungi.

The program was a rigorous one—individual instruction under senior teacher-researchers, work assignments in the several departments cooperating in the program, and designing and conducting their own experiments in the research laboratory. All spent specified times in the diagnostic laboratory at Barnes Hospital, working on daily routine analyses and sharing responsibility for identifying fungus cultures taken in the dermatology clinics. They reported on their own research projects at weekly sessions attended by the senior investigators.

The heavy research emphasis was geared to the philosophy of Washington University's School of Medicine as set forth by Samuel B. Guze, vice chancellor for medical affairs, in a letter to Sam Smith at Research Corporation in 1973: "Major frustrations for patients and physicians alike inherent in current medical practice stem directly or indirectly from medical ignorance. Usually, if medical knowledge permits prevention or effective and definitive treatment, everyone is satisfied and costs are tolerable. Dissatisfactions of all kinds arise when prevention or definitive treatment are not possible. Chronic suffering and disability and high costs then ensue. Research offers the only hope for improving the situation, and first-rate people must be trained to do the research."

The research results of the Washington University program were substantial. Several of the studies supported by the Brown-Hazen grants centered on the fungus *Histoplasma capsulatum,* the agent responsible for histoplasmosis, a disease already noted as closely resembling tuberculosis. When mistakenly diagnosed as tuberculosis and treated as such, histoplasmosis not only does not respond to the prescribed medication, but may actually be made worse by the secondary effects of the drugs.

The location of the university and its hospitals was most appropriate for the *H. capsulatum* studies, for St. Louis is in the central Mississippi valley, one of the regions of the United States where histoplasmosis is most prevalent. Here, as in the case of the most common fungus diseases, physicians, hospitals, and laboratories in endemic areas are aware of the prevalence of a particular disease and knowledgeable in its diagnosis and treatment.

One *H. capsulatum* study at Washington University was that of the reversible transition of this unique organism from the single-cell yeast stage to the multicellular mycelial stage as the result of temperature changes. In the multicellular state, *H. capsulatum* is a white, fluffy mold which, when disturbed from its normal habitat in the soil, becomes airborne. Breathed into the lungs, the spores go back to the dangerous single-cell state and proliferate in the favorable environment, sometimes becoming life-threatening. The university research team theorized that during the transitions of *H. capsulatum* from the single-cell to the multicellular state mutants might be produced which would be incapable of further change. Since mutants are often nonvirulent, the investigators hope that one or more may be isolated for use as a vaccine to prevent the disease. Such a preventive for the populations living in the endemic histoplasmosis regions could be a valuable public health safeguard. Whether or not a successful vaccine is ever developed, there is now a better fundamental understanding of the chemistry involved in the *H. capsulatum* transitions.

Another continuing study at Washington University is an attempt to find better methods of treating patients with fungal infections. Amphotericin B, an antifungal agent that has proved to be extremely valuable in arresting the most serious fungus infections, can have quite harmful side effects, including kidney damage. Ways are being sought to decrease the dosage of amphotericin B and increase its effectiveness, thereby achieving a real gain in fungus disease therapy. A largely unexplored area in which the research team is working is the possible effect of amphotericin B on the immune system of the human body. Since this is a drug whose chemical structure is well established, it serves as an agent of known characteristics for the study of the mechanism by which the immune system may be activated. The researchers feel that the antifungal agent may by some means trigger production of antibodies to battle the invading fungus. If this is so, it may be the immune response, not the antifungal drug, that is effective against the infection; or it may be a combination of the two. Resolution of the question may lead to a completely different strategy in the use of amphotericin B in treating fungus diseases.

A finding from another research project also aimed at improved therapy for fungus diseases may lead to more effectual chemotherapy for cancer victims. From their studies of amphotericin B, the Washington University researchers found that the drug binds to cell membranes in a way that increases their permeability and opens them to the entrance of substances that otherwise would be held back or barred. Thus, other antifungal agents administered with amphotericin B may more readily enter fungus cells to destroy them than is the case when the other drugs are used alone. It has now been shown that combinations of lesser amounts of amphotericin B with other agents can control several types of fungi while reducing significantly the side effects normally associated with amphotericin B.

This work has been further extended to the drugs used in treatment of cancer, with the amphotericin B in effect opening the way for the anti-tumor drugs to enter cancerous cells. The findings are now being reviewed by the National Cancer Institute and similar work is going on at medical centers across the country. As in the case of the antifungal drugs, these combinations in cancer chemotherapy promise to provide more potent treatment while producing fewer undesirable side effects.

In their study of amphotericin B, Kobayashi, Medoff, and their associates were intrigued by the similar chemical structure of that drug and nystatin as shown in the illustration at the bottom of page 84. The differences between these two complex molecules are relatively minor, yet the two drugs have very dissimilar characteristics and very different uses in fungal therapy. Nystatin is not easily soluble, therefore is most useful for direct application to fungus infections on the skin and mucous membranes—such as the eye, throat, and intestinal and vaginal surfaces—that can be reached by oral administration or by the use of salves or ointments. Amphotericin B is more soluble, therefore can be carried in the blood to the parts of the body where it can attack the systemic mycoses, the infections caused by fungi that have penetrated deep into the body.

Interestingly, the development of amphotericin B—which threatened for a while to be a direct competitor of nystatin—was apparently hastened by the work Squibb did in the 1950s in bringing nystatin from the laboratory stage to production. Rachel Brown reported that Squibb researchers had earlier worked with what appeared to be a powerful antifungal agent, but that the work was suspended when immense difficulties were foreseen in obtaining necessary quantities for commercial use. She had been told that, as the development of nystatin proceeded, techniques used in that process were found adaptable to production of the other antifungal substance, and the project was pushed forward again. The result was amphotericin B, a potent weapon against some of the major fungal infections.

A result of the research projects at Washington University has been the dissemination through respected medical and scientific journals of research results that have added to the store of basic information on medical mycology now available to all workers in the field. The postgraduate trainees involved in the research have had the experience of working on significant problems with experienced senior investigators and have now turned their training to further research of their own and the teaching of others.

The first eleven men and women trained in the postgraduate program went forth to teaching and research posts in institutions from New York to California, from Texas to British Columbia, and one in Spain. Four have been appointed to the staffs of medical schools. Three are in departments of biology in universities. One is a research associate in a private hospital, one is head of bacteriology and mycology at a national laboratory, one is a physician at a Veterans Administration hospital, and one is in a department of pathology at a U.S. Air Force medical center. In addition, four fellows who had been aided by the Brown-Hazen grants were still in training at the end of the grant period.

Following expenditure of the last of the $118,000 granted from Brown-Hazen funds, Kobayashi and Medoff reported in 1977 that the grants had made possible the establishment of a postdoctoral training program in medical mycology which would be continued with a U.S. Public Health Service training grant. This followed the pattern of many other Research Corporation grants, in which the initial impetus of private foundation money has allowed investigators to advance their research and their professional careers to the point where their work could attract support from the larger governmental funding agencies.

The postdoctoral program at Washington University is now recognized as one of the outstanding medical mycology programs in the United States. With no more than three or four openings a year for new candidates with Ph.D. or M.D. training, it receives thirty to forty applications every year for the fellowship appointments. The impact the trainees will have on the fight against fungus diseases is yet to be established, but eleven additional people at eleven different institutions are now helping to overcome a major insufficiency in modern medicine—trained medical mycologists to do further research in understanding the fungus diseases and in finding better means of treatment, as well as to teach a new generation of researcher-clinicians in a long-neglected field.

The program at the University of Kentucky College of Medicine, also started in 1973, was designed to attack the problem of the fungus diseases by a very different route. Like Washington University, the Lexington, Kentucky, institution is in one of the major histoplasmosis belts of the United States and in a region that, to an even greater degree than St. Louis, is historically plagued by tuberculosis, which pulmonary histoplasmosis closely resembles.

The airborne spores of *H. capsulatum* are breathed without apparent damage by most people in areas where the fungus is known to be widespread. In some people, however, and for reasons still not fully understood, the pathogenic fungi find a hospitable environment in the lungs. Thriving there, they form lesions that calcify, leaving a pattern on an X-ray plate that is virtually indistinguishable from the calcium formations that mark tuberculosis damage.

In areas where tuberculosis is the better-known and more-expected disease, physicians and laboratory technicians may mistakenly interpret the calcification as evidence of tuberculosis. The treatment that follows has no effect on histoplasmosis, and the patient with that disease grows progressively worse. In the most severe cases, the untreated disease disseminates, spreading to other parts of the body, frequently beyond any possible treatment and leading to death.

The obvious answer to this problem—which exists with many other fungus diseases as well—is better and faster diagnosis, a responsibility split between the examining physician and the laboratory technologist. Since physicians traditionally receive little or no training in medical mycology while in medical school, and laboratory workers often have no opportunity to study pathogenic fungi in the course of their training, the Kentucky program was designed mainly for these groups, with particular emphasis on the laboratory technologists.

To reach these workers, who could not take long periods of time away from their jobs, the mycology program provided short-term, intensive, advanced training courses—five days or less. For physicians and veterinarians there were also short-term courses, adjusted to their special interests. The overall program also had a month-long summer course for supervising laboratory microbiologists and for academic microbiologists who were teaching or doing research in medical mycology. Finally, there were two-year postdoctoral fellowships for comprehensive academic and clinical studies to be awarded to a small number of Ph.D.s and M.D.s to enable them to go on to faculty positions where they could, in turn, establish further research and teaching programs.

Facilities available for the project included the mycology laboratories of the University of Kentucky Medical Center and the research laboratory in mycology at the Lexington Veterans Administration Hospital. Together they provided teaching, research, and patient services, including clinical consultations and laboratory diagnostic services in human and animal mycology. The Medical Center's laboratories also served as a reference laboratory for the six state tuberculosis hospitals operated by the Kentucky State Department of Health, and there was an affiliation with the Regional Veterans Administration Center for Sero-Immunology and Mycology of Fungus Diseases. This combination offered a group with expertise in the various facets of medical mycology, and a wealth and variety of patient and mycological material for training purposes.

Prior to applying for a Brown-Hazen grant in 1972, the directors of the program had tested one major aspect of it—a short summer course in laboratory techniques in mycology for graduate medical technologists. The enthusiasm of the participants and the requests of others for training of this kind—plus an invitation from Jack Powers, the Research Corporation grants representative for that area—led to a proposal for three-year grant support of a full program, including the two-year course for postdoctoral fellows.

As in the case of the Washington University proposal, the outside referees and the members of the Brown-Hazen Committee were impressed by the facilities available to the program and the caliber of the medical scientists who would supervise it. The project directors were Ernest W. Chick, who was professor of community medicine and medicine at the university and chief of the research laboratory in mycology at the VA hospital, and Michael L. Furcolow, professor of epidemiology in the Department of Community Medicine and the Department of Pediatrics at the College of Medicine. As consultants and guest lecturers the program had Howard W. Larsh, research professor of microbiology at the University of Oklahoma, Norman, and Norman F. Conant, professor emeritus of microbiology at Duke University College of Medicine, Durham, North Carolina, both eminent in the field of medical mycology.

The proposal was a difficult one for the Brown-Hazen Committee, for it had in it less research, as such, than had been hoped for in training programs. Yet it focused directly on an acknowledged problem—upgrading the skills of physicians and diagnostic laboratories. After a good deal of discussion the committee inclined favorably toward the proposed program, but at Brown's suggestion, voted to fund it for only one year, then to welcome further requests if progress reports showed reasonable gains being made. As the reports came in and as the outside referees continued to evaluate the ongoing program, there was evidence of real progress. Subsequent grants were then made, ultimately amounting to $132,000, to support the program through 1977.

During the course of the five years, Chick left the program to go with the Bureau of Health Services of the State of Kentucky, and Furcolow became emeritus professor but still served actively in the program. Direction of the effort passed to Norman L. Goodman, who had been project coordinator from the beginning; he held joint appointments as professor of community medicine at the university and research associate at the Veterans Administration hospital.

In the terminal report on the grant, Goodman noted that the program had been carried out much as planned, except that fewer postdoctoral fellowships had been awarded, the institution's review committee for the program having decided that some of the funds for that purpose could be more effectively used for the short courses for laboratory workers and physicians. In all, seven postdoctoral fellows were trained during the period of the Brown-Hazen grants, four supported by the grants and three by other funding. They received intensive instruction in mycology and serology at the clinical laboratory, and in epidemiology, as well as patient care and management. Each carried out a research project in medical mycology and took part in the short-course teaching program.

The four Brown-Hazen fellows went on to establish or augment programs in medical mycology in universities and affiliated hospitals—two in Kentucky, one in Ohio, and one in Texas—all being actively engaged in undergraduate and graduate teaching and research. The three supported by other funds received appointments, respectively, as head of a public health program in Uruguay, professor of microbiology at a university in Indonesia, and head of a mycology-microbacteriology laboratory at a major clinic in the United States. After exhaustion of the Brown-Hazen grants, the university obtained other funds to continue the fellowship program on a reduced scale, offering one fellowship per year.

The portion of the program beamed at physicians had been extremely well received at its start, with 60 attending the courses in 1973, 42 in 1974, and 23 in 1975, at which time it was decided that the courses not be held annually, but at two- to three-year intervals. The pool of interested physicians was being drained and their turnover rate was considerably less than that of the laboratory technicians.

The short-term courses for laboratory microbiologists had continued to draw large attendance throughout the period, although there was a reduction in the number that could be invited toward the end of the grant period as the funds were running out. These courses had been made short-term because of the difficulty laboratory personnel would have in leaving their jobs for longer periods; they were also given free of charge in the beginning to attract those whose laboratories had no funds for travel or continuing education of the technologists. When a small fee was charged later to help defray expenses, there was a change in the character of those attending: fewer bench workers and more supervisors.

Over the five-year period 418 people received advanced training of varying concentration in medical mycology, most of them in the short-term courses. They included 131 physicians; 3 veterinarians; and 284 laboratory technologists, supervisors, and academic microbiologists. Of the latter group, 40 held the Ph.D., 24 the M.S., and 220 the B.S. degree.

Taking the 1977 participants as reasonably representative of the five-year total, a year later 60 percent were doing diagnostic microbiology and teaching, 29 percent laboratory diagnosis only, and the remaining trainees of this group were working in research and clinical practice. Of the 51 participants in 1977, 36 came from outside Kentucky, representing 13 states.

On the assumption that they absorbed the training given them through the program, the participants took with them on returning to their various locations a basic expertise in laboratory diagnosis: the ability to collect a clinical specimen for use in isolating fungi, to select the correct procedure for processing the specimen, to use the proper culture medium for growing a given fungus, to identify tentatively the fungi causing major mycotic diseases, to use approved procedures for shipping clinical specimens or cultures to a reference laboratory, and to relay and interpret laboratory findings to the physician.

A survey conducted in 1978 brought responses from 113 former participants in the training program, nearly 80 percent of them currently working in hospital laboratories. More than four-fifths felt that the courses had increased their proficiency in identifying fungi, and nearly one-half reported that they had been able to increase the recovery of fungi from patient specimens, the increases ranging from 5 percent to 100 percent. Better aid in diagnosis of the fungus diseases had obviously been one of the benefits of the program. A longer-term effect was the increase in the number of course participants who were teaching others; 30 percent reported teaching medical mycology prior to attending the courses, and 60 percent after the training. Thus, as a result of this one program, at least 33 more people over a broad geographical area had not only gained new knowledge of the mycoses, but were spreading the knowledge to still others. The program continues today at the University of Kentucky Medical Center on an enrollment-fee basis.

The value of these improved diagnostic services to the local areas served by the laboratories is pointed up by an incident reported by Goodman. Shortly after returning home to Ohio, one of the 1978 technician trainees phoned the project coordinator to tell him excitedly that she had tentatively identified one case of histoplasmosis and one of coccidioidomycosis, another serious fungus disease, neither of which she would have detected without her training at Kentucky. As a result, two patients whose diseases previously might have been wrongly identified could now be treated promptly for the correctly diagnosed disease. According to Goodman, this experience can be multiplied many times: the more people who are trained to recognize the fungus diseases, the more cases are uncovered.

11. *Invention Repays Research*

"Invention repays research" is a phrase associated with Frederick Gardner Cottrell, a chemistry professor at the University of California early in this century. He created Research Corporation to develop his invention, a novel means of cleansing industrial gases, and to use the proceeds to support research projects of other scientists. Cottrell's belief was that an inventor profiting from his own successful research (which undoubtedly also drew upon research results of earlier investigators) had an obligation to repay research by supporting further investigations of future researchers.

This concept appealed strongly to Gilbert Dalldorf in 1950 when he arranged for Cottrell's foundation to handle the nystatin invention. He held the belief steadfastly, finally writing in his memoirs his regret that "few have followed Cottrell's precept in accepting full responsibility for the appropriate use of their discoveries or the direction of independent programs for the support of science."

The arrangement he worked out with Research Corporation did follow the precept. It returned to diverse phases of scientific investigation all the inventors' rewards and granted nearly two-thirds of all Brown-Hazen funds for independent research projects. The money for project research was most often used for special equipment and supplies, and for stipends to undergraduates and research fellows. These were necessities, the "start-up" items, that investigators—particularly those not yet established in their fields—found it hard to get from their own institutions or from federal or other outside funding agencies. Brown-Hazen grants of this kind supported the work of more than 600 investigators, yielded hundreds of publications reporting new findings, often led to career advancement for the young investigators, and almost always helped to train aspiring youngsters in the biological-medical sciences.

As the grants program in its last few years turned to supporting research exclusively in medical mycology, the principals and advisers realized that the necessarily short-term effort would probably have no immediate consequences, but the hope was that some research projects started with the grant funds would lead ultimately to new weapons for the fight against the fungus diseases. Surprisingly, some of these efforts began to pay off during the remainder of the program, and, not so unexpectedly, some research projects funded earlier in the program helped pave the way for these later developments.

In 1958 and 1960, well before the program turned exclusively to the combat of fungus diseases, grants had been made to McGill University in Montreal for the mycological research of Fritz Blank, a German-born Swiss-trained chemist who had gotten into the field through his work on the chemotherapy of fungus infections at a pharmaceutical house in Basel. A further Brown-Hazen grant for his studies was made after he moved to the Skin and Cancer Hospital at Temple University in Philadelphia in 1962. Thus his work was well known to the Brown-Hazen Committee when he applied again in 1970 for a grant to support a broad research and training program in medical mycology at the Philadelphia institution. A total of $240,000 was granted for the program over the next six years.

Several specific research projects were included in the program, but its principal goal was to train qualified people in medical mycology and to interest them in doing research in that area.

Its success would be judged not only by research results, but by the direction in which the trainees would go after their training.

Following his belief that fundamental science needed to be applied to mycology, Blank and his colleagues at the Skin and Cancer Hospital set out to isolate cellular components of pathogenic fungi in an attempt to determine their function in the disease process. Their research emphasized the importance of purification and chemical characterization of antigens. They were successful in isolating a number of highly specific antigens appearing to have usefulness in identifying fungus infections, with the aim of developing improved diagnostic techniques and, eventually, new agents for treatment and prevention. In the training program, Blank designed and supervised one- and two-year terms for postgraduates pursuing careers in medical mycology.

In March 1977 Blank reported briefly to Research Corporation on the research and the training aspects of the program, noting progress made and adding that he would not fail to send a terminal report covering the last year of the grant. Unhappily, the promise could not be fulfilled, for he died suddenly the following month and the task of writing the terminal report fell to Sarah F. Grappel, who had worked closely with him on the program. Blank's death and the cessation of the Brown-Hazen support effectively ended the training program in medical mycology at the Skin and Cancer Hospital, but some of its effects are shown in the later careers of two of his trainees.

One is Roy L. Hopfer, assistant microbiologist at the M. D. Anderson Hospital and Tumor Institute in Houston, a unit of the University of Texas System Cancer Center. While working with Blank at Temple, Hopfer collaborated with his mentor on studies leading to development of a medium which provides faster recognition of *Cryptococcus neoformans*, the fungus responsible for cryptococcosis. This is an infection generally starting in the lungs from inhaled fungi, then spreading to other parts of the body, particularly the brain; in its disseminated form it carries a high risk of death. According to Conant et al. in their *Manual of Clinical Mycology,* "Cryptococcosis has been reported from every corner of the world where laboratory facilities have been adequate enough to be of aid to the clinic."[1]

Following his work with Blank, Hopfer's further investigations at the Houston institution led to a still more rapid test for *C. neoformans* which provides definite identification within six hours as opposed to three to seven days by other means. Fast identification of cryptococcal infection, either pulmonary or meningeal, is of particular importance in a cancer hospital such as his where the question always arises as to whether the patient's lesion is tumor or fungal. Quick and accurate diagnosis of *C. neoformans* allows prompt and correct treatment to be prescribed for the patient.

Commenting on Blank's program at Temple, Hopfer said recently, "I had never taken any courses in mycology, so it was an entirely new and exciting area for me to study. Since the training program, I have become and consider myself to be a 'medical mycologist.'" In addition to carrying forward his own research, Hopfer teaches medical mycology at the University of Texas Medical School and two allied institutions, where he is actively "spreading the word" about the importance of fungal infections to others in the medical field.

Another of Blank's trainees is Richard A. Calderone, assistant professor of microbiology at Georgetown University, Washington, who, subsequent to his work at Temple, was successful in getting Brown-Hazen project grants in 1975 and 1976 for his own research. One of his projects involved a study of the effect of rabbit-derived macrophages—cells that engulf and consume foreign materials—on two pathogenic fungi, *Candida albicans* and *Histoplasma capsulatum.* Fractions extracted and purified from the macrophages were shown to inhibit the growth of both fungi, apparently by slowing the transport of certain amino acids from the medium into the fungi. The extracts were able to kill *Candida* within two days.

Another of Calderone's projects, this funded by the National Institutes of Health, was the study of the disease process in *Candida albicans* endocarditis, a fungus infection affecting the lining of the heart and its valves, particularly after trauma which might occur in cardiac surgery. To date he and his associates have succeeded in microscopically characterizing the manner in which *Candida* interacts with damaged heart-valve tissue and how it survives on the valve by resisting the body's immune response. Unraveling the mechanism by which this occurs could throw new light on means of coping with not only *C. albicans* endocarditis but other types of fungal endocarditis.

Like Hopfer, Calderone is spreading the word about the importance of fungal infections through his teaching. He works routinely with medical students, giving them the opportunity to deal directly with pathogenic fungi in the laboratory, indoctrinating them to be alert to the possibility of fungus disease while making diagnostic judgments. In addition, he has taught two Ph.D. trainees, involving them in his own research, with both then having gone on to other medical schools where they are working in medical mycology, one at the University of Connecticut and one at the University of California, Irvine. Thus new links continue to be forged in the training-learning-training chain.

Another research project in medical mycology that may lead to a new weapon against fungus disease was started with Brown-Hazen funds at the University of Michigan Medical Center in 1974. It is being continued today at the Dermatology Department of Emory University in Atlanta with the university's own funds and grants from pharmaceutical firms. It began with a grant application from Henry E. Jones, assistant professor of dermatology at the Michigan institution, who requested funds for a study, "Host Resistance Mechanism in the Cutaneous Mycoses," dealing with the immune response of the patient to fungus diseases of the skin.

These infections, the dermatophytoses—including athlete's

foot, ringworm, and various tineas—are among the most common fungus diseases. They are caused by the dermatophytes, some thirty or more species of pathogenic fungi which invade only the superficial skin, in contrast with those which infect the subcutaneous tissues, bones, lungs, and other organs of the body. The National Health Survey of 1971–1974 projected from its sampling that nearly 16 million people in the United States had one of the dermatophytoses, with men four times as likely as women to have contracted the infections. The costs of these diseases in terms of physicians' services, medication, time lost from work, and productivity can only be guessed at. Medications alone run into tens of millions of dollars.

The Jones grant application in 1974 proposed an interesting approach to study of these diseases, noting that the subject of iron metabolism as related to fungus infections of the skin had not been seriously studied. Jones, who held an M.D. from Tulane University, requested funds for equipment and supplies essential to such an inquiry, and for a stipend to pay a postdoctoral student—William M. Artis, who had just received his Ph.D. in cellular immunology from the University of Wisconsin. A grant of $33,000 was made in 1974 and, upon demonstration of progress, a renewal of $40,000 was approved in 1976.

Today, under the leadership of Jones, the dermatology department at Emory is emerging as a major referral center for complicated skin disease problems in the southeastern region of the country. Jones is professor of dermatology as well as department chairman, and Artis, now assistant professor of dermatology, is director of the laboratories. Their research, interrupted by the move from Michigan, was set in motion at Emory with the help of the Brown-Hazen funds, also transferred from Michigan, as was some of their most sophisticated instrumentation.

The Jones-Artis research, like much other work in the sciences, has been significantly aided by various new technologies.

Their project required knowledge of changes in growth rates of fungi so they might determine the effects of giving—or withholding—substances they believed would affect growth, thus leading to the most likely avenues for controlling or destroying the fungi. To accomplish this, they adapted for the study of fungi an automated radiometric microassay method developed for immunologic investigations.

Using radioactively tagged nutrients, they found that growth rates—which they had established by conventional means—correlated closely with the uptake of radioactivity by the fungi. Then, by measuring the radioactivity of the fungi, they were able to test quickly the effects of variables in the nutrients, condensing months of experimentation into days. Among their findings was that iron is an essential micronutrient for dermatophytes; it followed, then, that if these fungi could be deprived of iron they could be destroyed.

The dermatophytic fungi generally colonize only the dead, superficial outer layers of human skin, where iron is plentiful, loosely bound, and therefore easily available for growth. The fungi grow inward from the surface until they reach the first of the living cell layers below. At this level, although iron is plentiful, it is not readily available because of the presence of the iron-binding protein, transferrin. The inability of the dermatophytes to obtain iron from living tissue explains why they do not cause systemic disease.

At the skin's surface, the body's cell-mediated immune system mounts an offensive which manifests itself as visible inflammation and almost always correlates with inhibition of the lateral spread of the infection. The studies of Jones and Artis suggest that mediators of this cell-mediated response (lymphocytes and lymphokine) do not directly inhibit fungal growth but rather disrupt the skin, allowing transferrin to percolate into the superficial outer layers where the fungus resides. There, transferrin stops fungal growth by depriving the fungus of iron.

Jones and Artis are now working with chemical compounds that may be introduced into the patient's body so as to further the process by which iron is denied to the fungi.

One result of their work is that the combination of immunology with mycology has already produced new knowledge as to the possibility of combating fungus diseases by manipulation of transferrin. A more tangible goal is the development and identification of compounds to deprive the dermatophytes and other fungi of the iron they need for growth, which, if successful, would provide drugs to treat patients having contracted the infections or to serve as preventives.

Also, their automated radioassay method of measuring minute changes in growth rates may be adapted by researchers in other fields for studies of other microorganisms and substances. In medical mycology, the method has great promise for efficiently testing various antifungal agents against specimens of fungi taken from patients, quickly identifying the agents that have the greatest activity against the particular invader. This would allow the proper therapeutic agent to be given to the patient at the earliest possible moment, increasing the chances of stopping the infection.

Meanwhile, on the opposite side of the continent, Antonino Catanzaro has been seeking answers to some of the mysteries of coccidioidomycosis, one of the deep, or systemic mycoses, an infection contracted in mild form by millions of people in the arid southwestern states and particularly in the San Joaquin Valley of California. *Coccidioides immitis,* the fungus causing the disease, flourishes in the soils of these areas and, picked up by the wind, enters the lungs with dust.

Most people contract a benign, self-limiting infection with symptoms of cough, fever, and chest pain which in time disappear without treatment. Others suffer from a progressive form of the disease that spreads from the original site and is disseminated throughout the body, attacking skin, bones, and vital organs. At this stage, the disease is highly malignant, with a death rate of up to 50 percent if not treated. Yet, no wholly satisfactory drug exists for its treatment, and the toxicity of amphotericin B, widely used to treat the systemic mycoses, severely limits the amounts of that drug that can be given to patients.

While the search for better antifungal drugs goes on, Catanzaro, who is assistant professor of medicine at the University of California Medical Center in San Diego, is using another approach—a large-scale experiment to evaluate what appears to be a promising means of manipulating the immune system to combat coccidioidomycosis.

It is the immune response of the body to the fungal invaders that determines why only relatively few of the many people infected develop the life-threatening form of the disease. Patients at this severe stage have impaired immunity, which can in some cases be rebuilt, with a consequent hastening of recovery. One agent used for treatment is transfer factor, an extract of human leukocytes, or white cells, which is obtained from the blood serum of donors who are not infected with the disease. However, in the scattered cases previously reported, the results of treatment with transfer factor had been beclouded by a number of uncontrolled variables. The need for far more reliable data was pointed up by an editorial in the *1974 Year Book of Medicine*, which noted that these "anecdotal" reports might have enough substance to justify some randomized double-blind clinical trials. "If such an approach is not taken," it continued, "we might find ourselves wallowing in undue optimism or pessimism with no firm data to justify either accepting or abandoning transfer factor as a reasonable mode of therapy."[2]

It was just such a randomized double-blind experiment that Catanzaro proposed to the Brown-Hazen program. It was to be a project of huge proportions, and very costly—more appropriate for federal funding than for a private agency. But in the absence of other funds for what appeared to be a most useful study, the Brown-Hazen Committee voted an initial $54,000 in

1975, followed by $47,000 the next year. Funds for the project were also obtained from other sources.

Catanzaro enlisted the help of a group of investigators in California, Arizona, and Texas—physicians in areas where coccidioidomycosis was most prevalent—who would follow prescribed protocols in randomly selecting, then treating and following up patients in the disseminated stage of the disease. Hundreds of people would need to be tested in choosing healthy blood donors. From their blood samples transfer factor would be separated, prepared, and shipped to the participating doctors for administration to their patients. Arrangements would have to be made for travel, lodging, and hospitalization of patients brought to the Clinical Research Center in San Diego. The experiment was to be double-blind, in that neither patients nor physicians would know whether the doses being administered were placebos or active preparations.

Before embarking on the full experiment, Catanzaro, working with the physicians in the Coccidioidomycosis Cooperative Treatment Group which had been organized for the task, conducted preliminary trials. In one group of 49 patients who received transfer factor along with amphotericin B, 31 showed gradual improvement over periods of many weeks or months. While this group had not responded to amphotericin B alone, the researchers did not feel certain enough of the effects of transfer factor to conclude that it was solely responsible for the favorable outcome. However, among the 31 such cases, there were 12 in whom improvement was clearly associated with the administration of transfer factor—either when a dramatic change occurred promptly, or when patients whose course had been clearly downhill, with medical prognosis gloomy, began and sustained recovery.

The full-scale trials directed by Catanzaro were not yet concluded when this account was written, and while the outlook was good for transfer factor as a therapeutic agent, the researchers were loath to forecast its effectiveness on the basis of preliminary data. The only certainty is that a great deal more will be known about the possibilities of immunotherapy as a weapon against one of the most serious fungus diseases.

Invested in these researches and those of hundreds of other investigators is more than $4 million in nystatin royalties, a handsome repayment of research by invention, one that was a tremendous satisfaction to the donors and their mentor and would have gratified Cottrell had he been alive to witness this extension of his credo.

As to the results, the returns are not in. Research by definition is entirely unpredictable and there is no way of knowing now how much these and other Brown-Hazen-funded researches will contribute to the arsenal of weapons to fight the fungus diseases. Some look promising for early application, some may be years in development, others may never produce a public health benefit. But now a small trained research army is in the field, working in many different ways toward solutions of the unanswered problems. These and the new generations of researchers they train are the fighters who will carry on the battle.

12. Diversity of Support

Except that it operated in the biological sciences, the Brown-Hazen program had marked similarities to the already operating physical science program of Research Corporation, which was singularly free to experiment in new directions. Commenting on the latter in a study of the foundation's grants programs over a twenty-five-year period, Carl W. Borgmann, former head of the science division of the Ford Foundation, wrote in 1972: "Considering the modest funds that were made available for grants, Research Corporation's behavior has almost been quixotic. Few windmills were so large as to be safe from its lance."[1]

Not only were the guidelines for the Brown-Hazen program broadly drawn, but members of the advisory committee and foundation advisers alike were willing to consider a wide diversity of proposals. With the committee consisting usually of an organic chemist, a microbiologist, a biochemist, and three physicians, with specialties ranging from virology to pedagogy, it was seldom that at least one member was not a knowledgeable partisan of a given grant application. Even proposals falling completely outside the announced interests of the program received attention, and occasionally funding.

"These 'diversities,'" Dalldorf told me, "were an exercise in the pursuit of freedom and of independent judgment. They were relished most of all by those with long experience on official—usually federal—advisory committees." He recalled "with special delight" that small and unconventional opportunities might yield large rewards, "often enriching the sponsor as well as the state of our knowledge and the affairs of the earnest investigator."

The $4,114,000 expended by the Brown-Hazen program for research projects covered an enormous spectrum of interests, and the diversity extended to the other types of grants as well. Over the life of the program, a further $686,000 was granted for research and training grants in medical mycology at universities and medical centers, and $685,000 for long-term funding of fellowships and scholarships for medical mycology and for training women in the sciences. And $636,000 was given in the form of departmental grants in biology at liberal arts colleges. The remaining $588,000 was used to meet a wide variety of needs, including programs for teaching undergraduates and graduates in fields other than medical mycology, convening of seminars and symposia, preparing and distributing publications, and a host of other efforts to add to or disseminate knowledge of the biological sciences. A total of $6,709,000 was devoted to these purposes during the twenty-one-year program.

Not only did the grants cover many types of scientific work; they also spanned broad geographical areas. Between 1957 and 1978 these funds supported projects in forty-five of the U.S. states and Puerto Rico, in four Canadian provinces, and in six other nations of the Americas: Argentina, Brazil, Costa Rica, El Salvador, Honduras, and Mexico. Overseas, the grants assisted research projects in France, Kenya, Lebanon, Taiwan, Uganda, and the United Kingdom.

Cutting across the broad categories noted above and included in the grant totals were awards made to four institutions that had contributed importantly to the training of the discoverers of nystatin and in the development of the invention itself. Principal beneficiaries were the New York State Health Department's Division of Laboratories and Research, at which Hazen and

Brown had done their research, and Columbia University, where Hazen had earned her Ph.D. and later learned the mycology which prepared her to initiate the antifungal research. The undergraduate institutions attended by the two inventors also received contributions in the form of capital gifts for scholarships and fellowships when the Brown-Hazen Fund was being phased out in 1977 and 1978.

As was contemplated when Hazen, Brown, and Dalldorf set up the fund, particular attention was paid throughout the life of the grants program to the special needs of the Division. As a branch of the Department of Health it was supported by state funds, but there were times when these could not be used for certain purposes, among them attendance of staff members at international scientific conferences held outside the United States. There were also other related activities useful to the Division and its staff, the costs of which could not be squeezed out of the annual state budgets. Brown-Hazen contributions of $306,520, although only a minuscule fraction of the Division's total budgets for the period, made possible a number of these activities which otherwise could not have been undertaken.

Nystatin royalties supported research projects of staff members, funded a series of summer training sessions for college and high school students, made possible M.D. or Ph.D. training for aspiring young Division scientists, and paid not only for the travel of Albany personnel to international scientific conferences but for lecturers from other institutions to come to the Division to conduct seminars. Other grants helped underwrite the costs of two major medical and scientific conferences held in Albany, the purchase of an electron microscope for the Division, training programs for technologists from local laboratories in New York State, and publications.

The Brown-Hazen Lectures, always eagerly anticipated and well attended, brought to the Division authorities in biology, pharmacology, genetics, biochemistry, microbiology, and other health-related disciplines. Initiated in 1958 on an annual basis and held most years through 1978, they were funded first by the Brown-Hazen program and later by grants made by Research Corporation in honor of the two women scientists.

The Brown-Hazen influence at the Division spread well beyond the effects of the grants themselves. After the nystatin discovery, one of Brown's projects was the further purification of the antifungal agent, then being developed commercially by Squibb. An assistant on the purification work was Fred Rapp, a Brooklyn College graduate who was a junior bacteriologist in Brown's laboratory in 1952 and 1953.

Rapp, now professor and chairman of the Department of Microbiology at the Hershey Medical Center of Pennsylvania State University, wrote to me recently: "My experience in her laboratory, which left us lifelong friends, certainly altered my scientific career. . . . Rachel had high standards but always encouraged her younger associates in their endeavors. Her quiet supervision and tactic of letting us strike out on our own in the laboratory were instrumental in furthering our thirst to be 'independent,' which, translated, meant to work for a doctorate." In 1956, three years after his work in Brown's laboratory, Rapp received his M.S. at Union College, and in 1958 a Ph.D. in medical microbiology at the University of Southern California, then advanced through the years to the position he holds today.

Hazen's personal influence also continued to be felt. Among the Division staff members who were enabled to go on to postgraduate work through Brown-Hazen grants was Sally Kelly, a senior research scientist who had come to Albany in 1951 with a Ph.D. in botany from the University of Wisconsin. An accomplished biochemist, she wanted to expand her training into the field of medicine, and in 1958 applied to the Brown-Hazen program for a fellowship. Her application was approved and for four years she attended New York University while still maintaining her affiliation with the Division, receiving her M.D. in 1963. She later became a research physician at the Birth Defects Institute of the State Department of Health, where she applied her twofold training to particular advantage.

In her book *Biochemical Methods in Medical Genetics,* a manual for laboratory technicians and physicians, Kelly pointed out the advantages of the duo-disciplinary attack on inherited metabolic diseases. "The clinician sees the disease—the biochemist identifies the underlying cause and means of control." Dalldorf, in his foreword to the volume, elaborated: "Medicine is finding many new opportunities in the expanding field of heritable diseases where the combined attention of biochemists and physicians has yielded revolutionary advances. Doctor Kelly, who was highly competent in biochemistry before undertaking her medical studies, ideally expresses the advantages of this broad approach."[2]

As Kelly was contemplating the advanced medical training in 1958, she had consulted Hazen, who by then had moved to the Albany headquarters. Kelly recalled later in a letter to Dalldorf the encouragement given by Hazen when she outlined her plans: "E. H. was direct, to the point, and succinct—she gave me her clear blessing for the M.D. venture with the single sentence: 'That's good; you'll be better off.'"

Another beneficiary of Brown-Hazen funds was the College of Physicians and Surgeons of Columbia University, which received grants totaling $198,700 between 1963 and 1976. A 1965 grant was used to remodel and enlarge the newly assigned facilities for the Laboratory of Mycology, one of the few "bricks and mortar" grants made from the Brown-Hazen Fund. This was the unit in which Hazen began studies with Rhoda Benham in 1944 as she was preparing herself for the new mycology assignment with the Division. An additional grant was made in 1970 and 1972 for a research and training program in medical mycology directed by Margarita Silva-Hutner. By the end of 1976, two postdoctoral and three predoctoral trainees had been prepared for medical mycological careers in this country, Japan, and Venezuela, and sixteen research projects had been completed. The research included studies of several of the pathogenic fungi, and had among its end results new aids to the identification of dermatophytes and the fungi causing chromoblastomycosis, and the development of purified antigens for skin and serological tests of patients suspected of having contracted sporotrichosis. Other Brown-Hazen grants to Columbia supported research projects carried out by faculty and staff of the College of Physicians and Surgeons and helped purchase an electron microscope for Barnard College.

An example of the "diversities" supported by Brown-Hazen funds was one proposed to the committee in 1972. It was not within the guidelines of the program, but an application had been encouraged and it had received unusually high marks from the referees and the staff. Members of the committee were enthusiastic too, Brown commenting, "This is a worthy cause and most suitable for our support," and Hazen writing, "Everyone is in agreement on the value of the mycological section of the ATCC." They were referring to the American Type Culture Collection in Rockville, Maryland, which maintains, among other collections, the National Resource Center for Living Cultures of Health-Related Fungi.

ATCC is a nonprofit organization that collects, preserves, and distributes to the scientific community reference cultures of microorganisms, viruses, and animal and human cell lines. Among its accessions are the original strains of *Penicillium notatum* that Fleming used in his discovery of penicillin, and of *Streptomyces noursei*, the nystatin-yielding microorganism that Hazen registered. A qualified investigator wishing to work with either of these, or any of the thousands of other cultures there, may acquire a subculture upon payment of a small fee.

While ATCC had been receiving very substantial support from the National Institutes of Health and the National Science

Foundation, fungi were not high on the priority list. Its 1972 request to the Brown-Hazen program was to permit increasing its collection of fungi important to human health.

Increases in opportunistic fungus infections following the use of antibiotics and immunosuppressants for patients suffering from other disabilities had led to new requests from researchers, physicians, and laboratory technicians for cultures of pathogenic fungi. These cultures, shipped by ATCC not only within the United States but to many other nations, provided a rich resource for research, comparative studies, and diagnosis of fungus diseases, and for training of students, doctors, and paramedical and technical personnel.

With the wholehearted endorsement of the outside referees who reviewed the proposal added to the Brown-Hazen Committee's own enthusiasm, a series of grants totaling over $94,000 was made from 1972 to 1976 to enlarge ATCC's collection of fungi. Over the period covered by the grants, well over 500 new strains and nearly 200 species were added to the collection of fungi pathogenic to animals and man—more than doubling the 1972 stock—and some 5,000 subcultures were distributed to scientists, physicians, and technologists. In his terminal report to Research Corporation in 1977, following expenditure of the last of the grant funds, Shung C. Jong, head of ATCC's mycology department and curator of the collection of fungi, wrote, "With the Brown-Hazen grant-in-aid, the American Type Culture Collection now maintains a most comprehensive collection of medically related fungi in the world."

Nystatin royalties reached their peak in 1974, the year in which the patent expired, with over $2 million being paid to Research Corporation, one-half of which was applied to the Brown-Hazen program, the other half to the foundation's grants programs in the natural sciences. With little more to be expected from royalties, the Brown-Hazen Committee began in 1975 to implement its plan for closing out the program, appointing an ad hoc committee to work out and recommend the specific allocations. Currier McEwen, the public member of the advisory committee, and James S. Coles, president of Research Corporation, were charged with this task.

The membership of the Brown-Hazen Committee had changed further by this time, with Morris Gordon, director of the Laboratories for Mycology at the Division, serving in Hazen's place and Robert P. Whalen replacing Hollis Ingraham. Gordon, co-author with Hazen and Reed of the third edition of *Laboratory Identification of Pathogenic Fungi Simplified*, had been named by Hazen as her alternate in the event of her being unable to serve; when she became ill in 1973 he substituted for her and after her death in 1975 he became a full member. Whalen had succeeded Ingraham as New York State Commissioner of Health and in 1975 assumed the commissioner's role on the Brown-Hazen Committee.

It had long been felt by a majority of the committee, the two women scientists being the dissenting minority, that the influence of Brown's and Hazen's work should continue beyond the termination of the grants program and that the means for this continuance should be scholarships named for them at their respective undergraduate institutions. McEwen and Coles worked out tentative arrangements with Mount Holyoke and Mississippi University for Women, suggesting further that the two institutions should to the extent feasible be given equal treatment.

One grant of $50,000, already mentioned, had been made to Mount Holyoke toward the purchase of an electron microscope for research and for instruction of undergraduates in the biological sciences. To match it, a grant of $50,000 was made in 1976 to Mississippi University for Women to buy equipment for one of its laboratories, which the university later named the Elizabeth Lee Hazen Microbiology Laboratories.

Equalization of the scholarship grants for the two institutions was somewhat more involved. In her will, Hazen had contributed directly to her alma mater, bequeathing to it a one-half interest in The Hazen Place in Mississippi, the proceeds to be used for a scholarship fund to aid girls attending the university. It was not until 1977 that the farm was sold at $1,200 an acre (Hazen's grandfather had probably paid no more than 50 cents an acre when he bought it before the Civil War), with the university's half amounting to about $185,000. The direct bequest was then augmented by a gift from the Brown-Hazen Fund of $115,000, bringing the total for Hazen scholarships at the Mississippi institution to $300,000. Mount Holyoke was awarded a total of $300,000 from Brown-Hazen funds for Brown fellowships in honor of its alumna. At both institutions the awards go toward the education of women in the sciences.

As of 1979, two Hazen Scholars had graduated from Mississippi University for Women, one of whom was working as a microbiologist in a Louisiana hospital, the other training in medical technology at a hospital in Tennessee. A Hazen Senior Award was made in 1979 to the outstanding senior science major, who enrolled in the Medical School of the University of South Alabama at Mobile.

The 1977 Brown Fellow at Mount Holyoke entered a Ph.D. program in organic chemistry at Stanford University; the 1978 awardee was working toward an M.D. and a career in geriatrics at the University of Rochester School of Medicine and Dentistry; and the 1979 recipient of the fellowship was one of five candidates accepted by Albert Einstein College of Medicine in New York that year for its seven-year program leading to the M.D. and Ph.D. degrees.

Along with the scholarship/fellowship programs at the two women's institutions, McEwen and Coles had recommended a fellowship program named for Gilbert Dalldorf and directed to postgraduate research and study in medical mycology. This followed wishes expressed earlier by both Hazen and Brown that Dalldorf be recognized for his part in their work, and also served as continuance on a smaller scale of the main thrust of the Brown-Hazen program in its final years.

Searching for an agency to administer the medical mycology fellowship well into the future, the committee chose the Infectious Diseases Society of America. In 1977, the IDSA established the Gilbert Dalldorf Fellowship in Medical Mycology, and $270,000 was set aside from the Brown-Hazen Fund to endow it. A year later the first fellowship was awarded.

The first Dalldorf Fellow, Thomas M. Kerkering, took up the fellowship at Medical College of Virginia, Virginia Commonwealth University, Richmond, from which he had earlier received his M.D. His sponsor and mentor was Smith Shadomy, professor of medicine and microbiology, and he was also under the tutelage of H. Jean Shadomy, professor of microbiology and medicine. Both members of this husband and wife team were active in separate researches on pathogenic fungi, as well as in teaching medical students, graduate students, and Infectious Disease Fellows. Both had received Brown-Hazen research grants in 1974, which they credited with helping to keep alive the medical mycology program in which Kerkering enrolled.

The Division of Infectious Diseases of the Medical College of Virginia had from its beginning in 1965 principal clinical and research interests focused on medical mycology. Initially it was supported by training and research grants from the National Institutes of Health, but upon their termination there was a vacuum in both support for and clinical interest in medical mycology. The earlier Brown-Hazen grants had permitted the Shadomys to maintain their research activities and their training of students, and were at least in part responsible for the granting of a total of five master's degrees and two Ph.D.s in medical mycology, as well as for aiding three postdoctoral students. "All of these endeavors," Smith Shadomy wrote to me, "helped to maintain the research environment which would serve to support Dr. Kerkering as the first Dalldorf Fellow. . . . If these interests had not been maintained, there would have been neither reason nor climate to support Dr. Kerkering's application for the Dalldorf Fellowship."

In collaboration with Smith Shadomy, Kerkering is developing a simple diagnostic test to correctly and quickly identify *Candida albicans* infections that have become disseminated. Cancer, leukemia, and transplant patients receiving chemotherapy or immunosuppressive therapy, and others chronically debilitated, are particularly susceptible to these secondary infections, which are potentially lethal when they begin to spread throughout the body. Kerkering and Smith Shadomy aim to devise a test for early detection of these disseminated infections so that antifungal therapy can be started when there is a better chance for cure and survival. They are well on their way, according to their latest reports.

The grants for the Dalldorf, Hazen, and Brown fellowships were the final ones made from the Brown-Hazen Fund as it ceased operations in 1978. While the program had gone through several phases before settling firmly on medical mycology, and had supported a multitude of different activities within the biological sciences, its main thread of continuity and its largest area of expenditure had been support of research projects. Through these grants the principal investigators—most of them in their early years as teacher-researchers—were enabled to pursue their independently conceived experimental projects. Their research results, reported in hundreds of papers published in scientific journals, have added previously unavailable information to the storehouse on which they and future researchers can draw for further work. And the thousand or more students whom they have involved in their research now constitute a motivated human resource to be applied to still unsolved problems.

Currier McEwen, the public member of the Brown-Hazen Committee and probably the most objective firsthand observer of its grants program, told me that he saw this as the most lasting contribution: "I was so accustomed to the NIH and large foundations' pattern of making grants—and large ones—almost exclusively to the horses out in front and bound to win that it was very refreshing to me to see the stimulus and help the Brown-Hazen grants gave to essentially unknown young scientists."

A more direct contribution to human health is in medical mycology, where for several years the Brown-Hazen program was the single largest nongovernmental supporter of research and training. Today a patient suffering from a fungus disease, superficial or serious, has a somewhat better chance than in the past of having the affliction correctly diagnosed and treated. Better and quicker ways of identifying the invading microorganisms, some of which have been described in this book, will lead to earlier and more accurate interpretation of symptoms and to application of the most effective therapy. And somewhere along the research and development pipeline are new antifungal agents and improved methods of using existing ones; also perhaps vaccines to immunize those most susceptible to disease-causing fungi. But a great deal still remains to be done.

13. Nystatin in the News

Nystatin proved to be a durable source of news stories appearing over the years and telling not only of its use in treating humans for fungus diseases, but of other completely unexpected applications that developed later. One particular newspaper story remained vivid in Rachel Brown's memory; it had come to her attention in 1957 as she was sitting on the steps of a dormitory at Massachusetts Institute of Technology, where she was taking a summer course in spectroscopy.

Always an early riser and for thirty years the one to open her laboratory at the Division in the morning, she was waiting for the MIT laboratory to open. Glancing over the *New York Times* of that day, June 29, her eye was caught by a photograph of Elizabeth Hazen and herself at the Albany lab, and the accompanying story, which was headed "Antibiotic Said to Be Effective against Some Fungus Diseases." It was the story of the issuance of Patent No. 2,797,183 to Hazen and Brown, and their assignment of it to Research Corporation.

It had been a long wait from the filing of the patent application over six years earlier, and while the inventors had been kept advised of its progress through the Patent Office, this was Brown's first notice that the patent actually had been issued. Bursting with pride, she looked up to see a few passersby paying no attention to her. "I can remember sitting there," she recalled, "with several strangers around, and saying to myself, 'They don't know I'm in the *Times*!' "

Earlier that year, nystatin had been in the international news in a very different connotation. A United Press dispatch dated February 23 stated: "President Eisenhower has sent ailing President Theodor Heuss of West Germany some medicine—mycostatin—by air together with 'heartfelt wishes' for his speedy recovery." International News Service added in its wire to newspapers: "The drug, mycostatien [*sic*], was obtained from the Naval Hospital at St. Albans, L.I., yesterday and flown to Germany aboard a Lufthansa Airline plane." In its story, the *New York Times* said: "The shipment, dispatched by the White House on the request of Dr. Heuss's physicians, arrived by air at Frankfurt this afternoon and was brought to Bonn by automobile by a United States Embassy courier. The drug, previously untried in West Germany, was identified by the trade name, Mycostatien [*sic*]."

In the margin of the clipping of the *Times* story Hazen preserved, she corrected the misspelling (only one of many that occurred over the years), writing, "Mycostatin is the Squibb name for our nystatin," and "We are, of course, excited by this." Despite President Eisenhower's good intentions, however, it is unlikely that nystatin would have helped the German president. His ailment was described as inflammation in the lower part of the right lung—a site that could not ordinarily be reached by nystatin, if indeed the infection were fungal. Nevertheless, the act and the story were typical of the overenthusiasm greeting the new "wonder drugs" that were coming onto the market.

Unlike some of the other new drugs, nystatin—when used for its prescribed purposes—did live up to expectations. Continued experimentation by physicians as reported in professional journals seemed only to confirm the first product claims made by Squibb. For ten years or more after the introduction of nystatin, papers reporting its effectiveness continued to appear in medical journals in the United States and abroad. It was credited with particular success in treatment of *Candida* infections of the skin, nails, and mucous membranes, including the mouth and intestinal and vaginal tracts. It was also cited as an effective agent in controlling secondary fungus infections in the intestinal tracts of patients being given broad-spectrum drugs.

Nystatin—although it was not known by that name at the time—had first appeared in the general news on October 17, 1950, after the Hazen-Brown discovery had been announced at the Schenectady meeting of the National Academy of Sciences. The item from the *New York Times* which Hazen clipped and saved reported that the antibiotic had been used so far on experimental animals only, but the results had been most encouraging, and that clinical trials would begin soon. Unnamed spokesmen from the New York State Department of Health were quoted as saying that the discovery "may have far-reaching effects, since none of the present antibiotic drugs have been effective in the treatment of fungus infections." Hazen, always supercautious in making claims for nystatin, had bracketed the words "may have far-reaching effects" and written opposite them in the margin of the clipping, "We did not say this."

Receipt of the Squibb Award in Chemotherapy by Brown and Hazen in 1955 resulted in a series of stories in the trade papers and the general press. In the story that appeared in *Knickerbocker News,* the Albany newspaper gave Hazen's birthplace as Clarksdale, Mississippi. The award recipient, in her usual practice of talking back to clippings, wrote in the margin: "This is a mistake, of course. I told the Squibb man that Clarksdale was the nearest town of any importance near where I had lived." The story also carried a photograph of Brown and Hazen in the Albany laboratory, the same picture that appeared later with the story of the granting of the patent. It was one that Brown grew resigned to as it cropped up again and again in other publications. "Oh, not *that* picture again!" was her usual response. According to one of her friends, however, the photograph (which is quite similar to the one used as the frontispiece of this book) did not actually displease her. The wave of publicity that had followed the nystatin discovery was so unexpected that it almost overwhelmed her, and her modesty required that she take no glory in it.

A different photograph—one of Hedy Lamarr—was featured in a 1965 article in the *New York Journal American* that cited Brown and Hazen, along with Lamarr, as among the very tiny percentage of women to whom U.S. patents had been issued. The account noted an accomplishment little known to the fans of the Hollywood star—her invention of an antijamming radio device which she patented shortly before World War II. Quoting a study of the U.S. Patent Office records made in 1923, the story stated that up to that time less than 2 percent of patents granted were to women. It added that in more recent years, including those in which Lamarr, Brown, and Hazen had patented their inventions, the figure had dropped to less than 1 percent.

Photographs figured again in a story carried by a number of newspapers in the Northeast in 1969 when Hazen and Brown were awarded honorary degrees in science by Hobart and William Smith Colleges in Geneva, New York. Accompanying the story in the *Geneva Times* were facing pictures of the awardees, one of which had news value in itself—that of Hazen. It was the first photograph in many years, and the last to the extent of anyone's knowledge, that she had posed for willingly. Brown never knew what impelled her colleague—always the despair of photographers—but Hazen had for some reason gone to a professional photographer for a portrait to be used for the occasion. Even stranger, she was so intensely proud of it that she chided Brown for the photograph she furnished the college to be published alongside hers. Hazen's was a studio portrait and Brown's one that showed the chemist member of the team in a workaday laboratory coat.

Hazen's temporarily reversed position recalled to Brown another occasion a number of years earlier, when her friend's aversion to cameras had already been established. It was a symposium on the therapy of fungus diseases held at the University of California at Los Angeles in 1955, at which Hazen and Brown were scheduled to give two papers. One, by Brown and Hazen, was to be read at a morning session, the other by Hazen and Brown, in the afternoon. Brown gave the paper on which her name appeared first, and during the luncheon break an enterprising photographer arranged to have the stage lights turned up full at the start of the afternoon session so he could catch Hazen unaware. To his disgust, it was Brown who came on stage to give that paper also, and the slides that were to illustrate her presentation were all but indistinguishable until she stopped her talk and called out for the lights to be dimmed. Although no one observed her, it is most likely that Hazen chuckled quietly as she sat in the darkened auditorium.

An article that appeared in 1967 in *Health News*, a publication of the New York State Health Department, opened with these sentences:

> The learned practitioner peered closely at the red rash that disfigured the elegant face. He motioned to an aide. Quickly, a shot of medication was applied. Almost at once the blotches disappeared.
>
> The priceless Renaissance fresco was as clear as it had been before the Arno River, in a flood of historic proportions, had drenched the art treasures of Florence. In gallery after gallery, the treatment was repeated. Frescoes and paintings alike were "cured."
>
> It was, reported *La Nazione*, a Florence daily paper, the first time such widespread treatment had been given mildewed paintings. The medication was nystatin, applied in a water spray with an air gun.[1]

The article continued with an account of one of the most publicized uses of nystatin for other than human ills—the restoration of art works after the Florence floods. In July of that year a major article in *National Geographic Magazine* covered the restoration, crediting nystatin for arresting the fungal growths and showing the restorers applying the antifungal material (see page 89). Six years later, noting the unveiling of some of the restored works, *Time* reported on the rescues made by the Restoration Laboratories in Florence, with the collaboration of major chemical companies, as art and science teamed up to save the centuries-old works of art.

For several years before the floods, organic chemists at the University of Florence had been experimenting with various antibiotics, including nystatin, to learn if their antifungal properties could be effective against mildewed paintings. One problem was that nystatin is not easily soluble and any remedy would have to be applied in liquid form. Eventually the researchers found a way to put the crystalline material into suspension in water for spraying onto art works.

When the Arno overflowed in 1966, the immediate flood damage was only too apparent, but after the waters had gone down, the damp paintings and other works were found to be further threatened by the growth of fungi. Umberto Baldini, head of the Restoration Laboratories, was quoted as saying, "If we had not found a solution, those frescoes would have been devoured by microorganisms." After trying dozens of mold-killing antibiotics for their effect on paint, the laboratory workers found one that would destroy the microorganisms without harming the pigments. *Time* identified this substance as "Squibb's Nystatin, a stomach medicine."

Still a different application of nystatin had been reported in the *East African Standard* in 1969 under the headline, "Drugs made for humans may cure coffee disease." Reporting on tests of compounds used to combat the fungus causing coffee berry disease, the Nairobi newspaper stated that although "a drug called *nystatin*" had been effective against the fungus under laboratory conditions, the experimenters had not yet been able to put the antifungal substance into a satisfactory form for spraying. Apparently unaware of the Florence experience, they held little hope that they could use it for treating trees in the coffee plantations.

Greater success was reported in 1973 in an article in the *Springfield Union.* Datelined Deerfield, a small community north of Springfield, Massachusetts, on the Connecticut River, the news story began, "Native and visiting celebrators of the town's 300th anniversary are finding the local scenery a little different from that in most other places in the Connecticut Valley, as several of the town's giant elms have been equipped with boxes on their trunks and funny little tubes in their roots."

Continuing, the story said that the unusual appendages to the trees were part of the Lowden Tree Company's effort to find a cure for Dutch elm disease, and that the firm was injecting the trees with an antibiotic called nystatin. Arthur C. Costonis, director of research and development at Lowden, was reported as saying that nystatin had been tried the previous year with a high success rate when used on newly infected trees, but that the drug did not give immunity, and that treated trees could contract the disease again, possibly in the same season.

Four years later, the *New York Times* of June 11, 1977, reported from Providence, Rhode Island: "A natural antibiotic often used to combat yeast infections in humans is also proving effective against a widespread blight devastating the Dutch Elm tree." The story noted that a four-year program using "the chemical Nystaten [*sic*]" had saved trees infected with the disease on the Rhode Island State House grounds. Describing the application process, the *Times* said that the chemical was fed much like a transfusion into the tree from containers attached to the trunk, and that the tree's circulatory system then carried the substance to afflicted areas (see photo on page 88).

This was a good example of delayed research payoffs, for a quarter of a century earlier when Hazen and Brown first reported the results of their research with nystatin, among the fungi successfully attacked by the antifungal substance was

Ceratocystis ulmi, the plant pathogen causing Dutch elm disease. Applying their finding and being aware of the difficulty of getting nystatin into solution, the Lowden firm found a solvent that would dissolve nystatin powder and not be harmful to the tree. Diluted to the proper concentrations, this liquid could be injected into the xylem of the tree for distribution to the infected parts.

The latter stages of testing nystatin for the plant disease also included experiments with elm trees on the eastern shore of Maryland, and involved the University of Maryland, Wye Tree Experts, Inc., of Wye Mills, Maryland, and the University of Massachusetts, Boston, as well as Lowden, Inc., of Needham Heights, Massachusetts. The project was aided in 1975 and 1976 by Brown-Hazen grants to the University of Maryland and the University of Massachusetts.

Reporting on the status of the treatment in 1979, Francis J. Townsend, head of the Insect and Disease Control Division of Lowden, stated that with proper controls, the infusion of nystatin was generally effective in treating trees minimally infected with Dutch elm disease, beginning to take effect as early as fifteen minutes after injection and lasting up to ten days. His firm was treating some 400 trees a year, finding that the drug had greater use as a preventive than as a therapeutic measure. While economically feasible in controlling the disease in elms of special historic or esthetic significance, the process is still too costly to use on large stands.

Not widely disseminated through the general press, but reported in specialized publications, were still more applications of nystatin, only some of which could Brown and Hazen have envisioned while doing their research. Among unexpected applications, in 1960 the Wine Institute of Bordeaux experimented with the antifungal material for wine preservation, and it was reported that similar uses were made by vintners in Italy and Argentina. Bananas from the Mediterranean destined for the United Kingdom were treated with nystatin for preservation during shipment, as were stalks of the fruit from Jamaica and Brazil. In the United States, after FDA approval in 1970, chickens and turkeys could be treated with nystatin in the form of feed additives for control of skin lesions and eye infections caused by fungi.

Among uses Brown and Hazen might have foreseen are those to which nystatin is being put by microbiologists, who find the antifungal material a valuable laboratory aid. Since it is well established which strains of fungi are resistant to nystatin and which are not, the substance serves as a genetic marker in helping to distinguish between similar specimens under study. Also, since it is an effective agent against several of the *Candida* strains—which are carried by most people and therefore are common laboratory contaminants—nystatin is applied to prevent such contamination in the growing of other cultures, particularly tissue cultures; if the fungi have already established themselves, it is used to destroy them before they obscure the experimental results.

With the exception of the Lamarr-Brown-Hazen story on women inventors, not many of the news accounts from the early 1950s to the late 1970s gave more than passing attention to the fact that the scientists whose work was being reported were women, or delved particularly into their personal characteristics and beliefs. One that did both, however, followed the presentation of the Chemical Pioneer Award to Brown and Hazen in Boston in 1975. The writer, David F. Salisbury of the *Christian Science Monitor*, was able to interview only Brown, since Hazen was too ill to attend the ceremony. His story, which appeared in the *Monitor* on October 1 and was later picked up by many other newspapers around the country, began: "If Rachel Brown had not been receiving a distinguished award for her pioneering work in chemistry, I would have thought her a kindly grandmother. That is how greatly she contradicts the stereotype of a successful career woman in the male-dominated fields of science."

Recounting the story of the Brown-Hazen research leading to

the discovery of nystatin, Salisbury wrote that in their work with the grants program supported by royalties on the invention they had made a particular effort to encourage and support other women scientists, but had been frustrated by a lack of response. He quoted Brown as saying that a few years earlier the program considered supporting more women, but after visiting a number of campuses, she found little interest. She added, pointedly, "You must make a choice and be willing to sacrifice some things for a scientific career." Her own life work, she told Salisbury, "has been very rewarding."

Elizabeth Hazen would have felt rewarded had she been able to take part in ceremonies announced by the *Commercial Dispatch* of Columbus, Mississippi, on April 15, 1977—the dedication of the new Elizabeth Lee Hazen Microbiology Laboratories at her alma mater, Mississippi University for Women. Brown, Dalldorf, and Coles were honored guests at the ceremonies the next day at which students, faculty, friends, and relatives of the microbiologist heard addresses by Brown and Dalldorf on the contributions of the 1910 graduate of the school.

Speaking at the dedication, Harry Sherman, head of biological sciences at M.U.W., summarized the progress made in that discipline at the college, noting that the grant made from the Brown-Hazen Fund the previous year "has placed our laboratories among the best-equipped undergraduate microbiology facilities in the country"—a far cry from those available to Hazen almost seventy years earlier at the Mississippi Industrial Institute and College, as M.U.W. was known then.

Hazen did have the satisfaction, some ten years before the M.U.W. ceremony, of helping to dedicate another laboratory which had also sprung from the work on nystatin. In May 1967, *The Stethoscope,* a Columbia University publication, reported the ceremonies opening a facility marked by a new bronze plaque which read, "The Hopkins-Benham Laboratory for Medical Mycology, a gift from the Brown-Hazen Fund of Research Corporation."

The name for the Columbia laboratories had been urged by members of the Brown-Hazen Committee to honor J. Gardner Hopkins, founder of the laboratory in which Hazen had received her first instruction and experience in mycology, and Rhoda Benham, who had taught and inspired Hazen during her training there. Hazen was able to enjoy the new quarters at Columbia for only a few years as she continued her research and teaching, consulting with her friend and colleague, Margarita Silva-Hutner, who was also Benham's student and later her successor as director of the Mycology Laboratories.

It is said by one who works in the Hopkins-Benham Laboratory today that there is still a pervading presence in the room, an unseen figure peering through microscopes, checking the media used for growing fungi, looking into lab notebooks, and silently exhorting, "You can do anything you make up your mind to do. Don't say you can't, because you can!"

14. *The Final Years*

The last meeting of the Brown-Hazen Committee that Elizabeth Hazen attended was held in Atlanta in May 1973. It was obvious on the second day of the meeting that she was ailing, so arrangements were made to hasten her return to New York and have her met at the airport. Her health had been deteriorating for some time and increasingly she had become a female version of the absent-minded professor, often misplacing her papers or her briefcase. Somehow she also developed a special knack of locking herself in hotel rooms so that outside help had to be called to free her.

Back in New York she regained strength enough to start work at Columbia again, but soon became extremely concerned about the health of her sister, Annis Harris. Annis was blind and incapacitated and, following the death of her husband, had been placed in a nursing home in Seattle, where the couple had lived for many years.

Elizabeth had been especially attentive to Annis, visiting her frequently and worrying about the kind of care she was getting. In July 1973, despite her own uncertain health, she flew to Seattle, planning to stay indefinitely, and had her sister moved to another nursing home, where she felt Annis would get better care. But as her own condition worsened, Elizabeth was hospitalized and then moved into the same home with her sister—

Mount St. Vincent Nursing Center. Annis died there in 1975 and Elizabeth became despondent, saying at one point that she'd like to take the elevator up to Heaven to join her sister.

Because Elizabeth was always so private about her personal affairs, her "sister-cousin," Clara Hazen, was concerned that she might not have enough money for the hospital and nursing home charges. After her retirement from the Division, she had refused to apply for Social Security, and—as far as Clara knew—had only the state pension and the income from the farm in Mississippi. Clara dipped into her own savings to transfer funds for her cousin's medical care.

Among those who crossed the country to see Elizabeth in the Seattle home were Beulah Townsley and her husband, Paul; Clara Hazen, Conway Dickey, Mildred Hearsey—and Rachel Brown. On the day Rachel first tried to visit her at Mount St. Vincent, Elizabeth was reported to have been disoriented, and she refused to see her colleague. Then, a day or so later, Clara Hazen and Beulah Townsley took her out of the home for a few hours and bought her a new dress. The next day when Rachel saw her and her two cousins, she was welcomed by Elizabeth, who was bright and chipper, possibly because she was wearing the new dress.

She steadfastly proclaimed that she would soon return to New York, her apartment, and her work at Columbia, but as her illness persisted her doctors had to tell her flatly that she could not. Rachel had accepted the Chemical Pioneer Award for her in May 1975 and Elizabeth had been told of the honor, but to bring the event closer to her a small reenactment of the presentation was planned at the nursing home. It had to be cancelled, for on June 24 as she was being prepared for dinner, she turned to the aide who was helping her, thanked her for her kindnesses and those of others at Mount St. Vincent, then slumped over in her chair. She was eighty-nine.

Her funeral services were private and she was buried in a Seattle cemetery, the last in a plot Annis Harris had bought for her husband, her sister, and herself. When Elizabeth's affairs were settled, it developed that she had not needed the money sent by Clara. She left, in addition to The Hazen Place in Mississippi, what was said to be a sizable estate, including some of the "blue chip" stocks which she had bought many years earlier with what she called her "play money."

In October 1975 a special memorial meeting of the Medical Mycological Society of New York was convened, bringing together three of her closest colleagues to speak of their friend. Dalldorf, Silva-Hutner, and Brown were thus able to commemorate her professional career before an appropriate audience. Their remarks and a collection of Hazen's scientific papers were bound and presented in her memory to Mississippi University for Women and to Columbia University. A further posthumous honor came in 1980 when Elizabeth Lee Hazen's biography was entered in the reference work *Notable American Women: The Modern Period,* edited by Barbara Sicherman and Carol Hurd.[1]

After Hazen's death, Morris Gordon served in her place on the Brown-Hazen Committee until its dissolution in 1978. Dalldorf continued as chairman and Brown as secretary through the final meeting at which commitments were made for the Hazen, Brown, and Dalldorf fellowships and scholarships. Brown recorded in the minutes of that meeting a letter from Charles Schauer, the original Research Corporation member of the committee. Schauer commented on the simultaneous coming to majority and demise of the Brown-Hazen program, noting that it was ending almost exactly twenty-one years from the day he had first visited Brown, Hazen, and Dalldorf in Albany to plan a grants program.

Shortly after the final committee meeting, Brown, accompanied by Coles, visited her alma mater, where she had the satisfaction of handing a Brown-Hazen check for $100,000 to David B. Truman, Mount Holyoke's president. It was a further installment toward the Rachel Brown Scholarship/Fellowship Fund, which had been inaugurated a year or so earlier. She was also understandably pleased at another incident that occurred while they were on the campus. She and Coles were concluding a long conversation with a freshman, the daughter of one of Coles's friends whom he had been asked to see while he was there. As they walked across the campus, Coles mentioned that they were headed for a luncheon at which the girls who has been chosen Rachel Brown Scholars could meet Brown; then he nodded toward Rachel. The girl stopped in her tracks and faced her, blurting, "Why, you're THE Rachel Brown! I've heard so much about you, but I'd never hoped to meet you."

Although her duties with the Brown-Hazen program were ended, Brown still kept busy in Albany, giving her informal talks on science and abstracting papers within her specialties for *Chemical Abstracts,* as she had done for over fifty years. She was a regular visitor at the Division headquarters to offer counsel and drop off copies of the scientific journals to which she subscribed. Dagmar Michalova, director of the library, said that these publications saved the Division thousands of dollars, but that Rachel's personal interest and counsel were even more appreciated.

In 1976 she had been elected to the vestry of St. Peter's church—the first woman member in its history—and in 1979 was made chairman of the Every Member Canvas and the church's capital fund drive. In that year she also led a fund drive of her class at Mount Holyoke to endow a 1920 room in the new Willits-Hallowell Alumnae-Faculty-Student Center, and was cited for fifty years of service to the Albany branch of the American Association of University Women, of which she had earlier been president. Another honor came in 1979 when her biography was included in *Women Pioneers of Science,* by Louis Haber.[2]

Rachel's eightieth birthday in 1978 coincided with Thanksgiving, generating a double celebration at 26 Buckingham Drive which was heavily influenced by the "Chinese connection." That day marked the high point of week-long festivities as various of Sing-mei's family came to deliver gifts and good wishes when they could get away from their jobs or other obligations. Rachel's own family was represented by Sumner and Ruth Brown. The home blossomed with Chinese decorations and on the birthday-holiday Rachel wore a red, high-necked morning gown bearing four symbols. The red of the dress, Sing-mei explained, meant happiness, and the characters translated into Happiness, Prosperity, Blessings, and Longevity. The celebrants agreed that these were indeed most appropriate to the honored person and the very special occasion.

Little more than a year later, after a period in which she carried on all her activities in fine health and high spirits, Rachel told Dorothy Wakerley one morning that she had had pains in her chest during the night. Two days later, on January 14, 1980, she died at age eighty-one in St. Peter's Hospital in Albany. At a memorial service held in St. Peter's Episcopal Church, the rector, Laman H. Bruner, Jr., recounted her contributions to the church, the Division, and to humankind. "She extended a profound influence on every area of life she touched," he said. "Our lives have been made richer and more knowledgeable because of her."

By coincidence, at about the time of her death another tribute came with the publication of the special January 1980 issue of *The Chemist,* journal of the American Institute of Chemists. Titled "The Future Role of Women in Science and the World," the issue signaled the accession to the presidency of the Institute of E. Janet Berry, chemist and patent attorney—the first woman to be so honored. Rachel was one of the contributors, women chosen from government, academia, industry, and other pursuits, who had demonstrated qualities of leadership and achievement.

In the article, her last publication, she reminisced on her sixty years as a chemist, noting that "the time is already past when an occasional woman scientist, such as Mme. Curie, was grudgingly recognized for her pioneering contributions to chemistry and physics." She closed with a thought reflected in her other writings: that it is imperative that the woman scientist "recognize her own potentials, believe in herself, and determine to give unstintingly of herself to her chosen career. Only then can she expect to achieve on a par with her male counterparts. No legislation can bring this to pass."[3]

Gilbert Dalldorf maintained his interest and activity in the Brown-Hazen program throughout its full life and after its formal termination. He was personally as well as professionally involved and was mainly responsible for shaping its policies, directing its operations, and surprising the other members of the committee with his sometimes unorthodox suggestions for funding certain "diversities," unusual proposals for scientific work that came to his attention.

He continued also to be somewhat uneasy about the way the program was being administered. From the start he had championed a rather freewheeling approach, so fearful was he of bureaucracy creeping in, as he had seen happen in other grant programs, principally those of the federal government. He voiced this succinctly in a letter to Brown as the program was just getting under way in the late 1950s: "I think our first rule should be not to fall victim to arbitrary rules—our own as well as others."

The granting policies Research Corporation applied to the Brown-Hazen program went in the same direction of nonrigidity as Gilbert's predilection, but it could not go as far as he wanted. Its actions were dictated not only by what it had found to be good granting practice in its other programs, but by state and federal regulations, particularly after the Tax Reform Act of 1969, which placed new restrictions on private foundations. The resulting Brown-Hazen program was a compromise, therefore, but one to which Gilbert eventually gave high marks in the last paper to appear over his name in a scientific publication.

In that paper in the *Journal of the American Medical Association* in 1976, Gilbert reviewed the program and what he saw as some of its accomplishments and shortcomings. He wrote in the conclusion: "Having served as chairman of the advisory committee, I believe that the best advice I have had came from successful investigators, and that personal acquaintance is very helpful, and at times essential to intelligent judgment."[4]

Three years later he had reason to reinforce these beliefs when he visited two successful investigators and made personal acquaintance with a younger researcher under their tutelage. The occasion was a trip he made in 1979 to Virginia Commonwealth University to meet Thomas Kerkering, the first Dalldorf Fellow in Medical Mycology, and Smith and Jean Shadomy, Kerkering's mentors. In their laboratories he saw and heard at firsthand what had been accomplished and what might be expected from their work toward early detection of disseminated *Candida albicans* infections. He was much heartened by the results of this application of Brown-Hazen funds and immensely pleased that his name would be associated with their work.

Two months later, as was their annual custom, Gilbert and Frances went to their daughter's vacation cabin in the Adirondacks for a family gathering. Gilbert had known for some months that his time was limited, but he had hoped to spend one more Christmas with the children and grandchildren. He failed rapidly, however, and died in his sleep on December 21, a few months short of his eightieth birthday.

More than a year earlier, when the malignancy was first detected, he had hastened to finish his memoirs, "Days of Our Years," which carried the subtitle "Recollections of a Golden Era of Science and Medicine and of the Men and Women Who Participated in It." Copies were sent to only a few friends and colleagues, carrying some thoughts he wanted to leave behind, among them a reminder of what he considered the salubrious climate of science in his younger days:

> In the beginning we were all amateur scientists. We were teachers or pathologists or biological chemists who were beset and fascinated by questions and problems that constantly arose in the

> wards and the morgue. . . . Curiosity and the fascination of advancing knowledge ruled. The material rewards were meager and often uncertain but the trade was respectable and stimulating. . . .
>
> Our system had other virtues. It provided so well for fledgling scientists who were eager to try their wings and who profited from these inconspicuous opportunities to venture small excursions and to learn by trying. It provided a high degree of freedom and freedom opens the door to intuition and to innovation. All were free to pursue truth in their own ways, regardless of the consequences.

Turning to a condition that concerned him greatly in his later years, he wrote:

> Independence is an essential for researchers and their enterprises. The huge federal programs that now call the tunes to which much of science dances interfere with this priceless freedom. Interference seems to be characteristic of government. Louis Pasteur was concerned on this score and understood it very well a hundred years ago, and it has been a source of distress to many of his followers ever since. It was, indeed, recognized by the men who designed our present practices and who undertook to restrain the evil with such devices as peer review and advisory committees from outside the federal services. Despite their good intentions, the burden of remote, central power has become even more evident and onerous as the programs have grown and management has proliferated.

In the closing section of his memoirs, Gilbert wrote of their boat and of Oxford, where Frances and he spent their last years together: "The Eastern Shore has proved to be a most pleasant haven. It is characteristically a land of good nature and friendliness. The waters invite sailing and fishing, and the camaraderie of old friends is a blessing."

15. *A Continuing Challenge*

When the Brown-Hazen program came to a close, it was acknowledged by those who had worked on it that only one of the two objectives announced for the program's final thrust against the fungus diseases had been met. There had been far greater concentration of funds on training to increase the numbers and competence of those working in medical mycology. The program had done this directly through the grants to several universities and medical centers for training and research.

The other objective—to try to attract additional funds and more people to the effort—had not been successful, particularly as to funds. With the departure of the Brown-Hazen program from the field, there was practically no private foundation or general public support for research and training in medical mycology, and there was no upsurge in federal funding.

Virtually the only federal source of funds for this kind of work at universities, hospitals, and other nongovernmental institutions is the National Institute of Allergy and Infectious Diseases (NIAID), under the U.S. Department of Health and Human Services, the former Department of Health, Education, and Welfare. However, both NIAID and the Center for Disease Control (CDC), also an agency of HHS, do conduct mycological research on an intramural basis, as does the Veterans Administration at individual VA hospitals. In the past, the VA and CDC supported cooperative research ventures, among them important studies on histoplasmosis and blastomycosis. Now, with these cooperative groups no longer in operation, a significant void has developed in direct federal funding of research on the fungus diseases, and NIAID can provide extramural funding only to the extent of its very limited budget.

NIAID tackled the problem in November 1977 when it convened the Workshop on Medical Mycology Research and Training to identify reasons for the decline in research support despite evidence that morbidity and mortality caused by fungus infections were an increasingly important medical problem. The Workshop was to recommend measures which could be implemented by the National Institutes of Health, of which NIAID is a branch. The first recommendation of the Workshop was the establishment of several national centers for medical mycology to be devoted to multidisciplinary science, be guaranteed support for five years, and have access to clinical cases for study. The Workshop urged that the centers have a strong teaching component in addition to research.

NIAID took action on the recommendation, making two grants in 1979 to fund centers for medical mycology, one at the University of California at Los Angeles, the other at Washington University in St. Louis, which had been aided earlier by one of the Brown-Hazen training grants. Reporting the grants to its members, the *ASM News* of the American Society for Microbiology commented, "The creation of these units reflects recognition, by NIAID, that fungal infections have become an increasingly important cause of disability and death in this country. Ironically, the emergence of this problem reflects the darker side of new treatments for malignant or immunological disorders; such treatments often appear to weaken the defense mechanisms that ordinarily prevent such infections."[1]

Also in accord with the recommendations of the Workshop, the NIAID grants to the two institutions provided five-year funding, supported multidisciplinary research, and placed the funds in situations where there is access to clinical cases. But a key component was missing: the funds could not be used for training, whereas the Workshop had urged that training funds be provided to guarantee a future generation of scientists trained in mycology.

Training was included, however, in three other five-year programs of lesser magnitude funded in 1979 by NIAID at Temple, Tulane, and Washington universities; these grants provided National Research Service Awards for predoctoral and postdoctoral research training in specified personnel-shortage areas. In 1979 also there was one NIAID Modified Research Career Development Award in medical mycology to Boston University, this type of support being given to foster the development of young scientists with potential for careers of independent research. Completing the list, there were thirty NIAID grants for individual research projects in medical mycology at universities and hospitals. The project grants accounted for $1.9 million of the total of $2.8 million committed in 1979 for NIAID support of medical mycology outside government laboratories.

NIAID has limited funds for its Mycology Program, for that program has to compete with the other missions assigned to the larger Microbiology and Infectious Disease Program of which it is a part—among them virology, bacterial diseases, and vaccine work. In 1979, mycology was allocated about 1.3 percent of the total NIAID budget.

NIAID in turn has to compete for funds with the other institutes within NIH, including the National Cancer Institute, the National Heart and Lung Institute, the National Institute of Arthritis, Metabolism, and Digestive Diseases, and the National Institute of Neurological and Communicative Disorders and Stroke. In 1979, NIAID was given slightly more than 6 percent of the total NIH budget. One and one-third percent of 6 percent could be taken as a measure of the ranking of the fungus diseases as a perceived national public health problem.

Yet the reports from the field continue to point to the severity of the problem. An epidemic of coccidioidomycosis in 1978 in an area where the disease rarely occurred was reported in the *New England Journal of Medicine* in 1979.[2] The cause was a dust storm late in 1977 over a part of the San Joaquin Valley of California where the disease is endemic. Unusual winds raised particles of soil and fungal spores, carrying them in thick clouds as far north as Sacramento, some 300 miles away, where they were deposited on the earth and in the lungs of a susceptible population.

In the first sixteen weeks of 1978, 550 cases of coccidioidomycosis were reported to the California Department of Health, far above the maximum of 175 for the whole state for that period of the year in any of the ten preceding years. Sacramento County alone reported 139 cases, compared with from none to 6 cases per year over the previous 20 years. Of the 139 cases, 115 met the criteria of the authors of the *Journal* report for dust-storm-related coccidioidomycoses. Eight of the 115 died, with only 2 of the deaths not attributable to the disease. The cost of medical care for the state-wide victims of the epidemic was estimated at more than $1 million.

In an editorial in the same issue of the *New England Journal of Medicine,* David J. Drutz of the University of Texas Health Science Center in San Antonio cited an outbreak of another fungus disease in a place also outside known endemic areas. At about the same time as the Sacramento epidemic, Indianapolis had an outbreak of nearly 350 clinical cases of acute pulmonary histoplasmosis, with 36 instances of dissemination and 14 deaths.

In Indianapolis, the source of infection was thought to be the disturbance of earth caused by construction, since most cases in the histoplasmosis epidemic were reported from sites downwind from that activity. As Drutz pointed out in his editorial, "The likelihood of acquiring histoplasmosis as an airborne 'pollutant' rather than through contact with a specific locus of fungal contamination is well known to mycologists. The common denominator has been disturbance of the environment, often by human activities, with subsequent airborne spread of fungi."[3]

Also reported in 1979 was the relative ineffectiveness of present-day treatment of a form of aspergillosis, a disease caused by one of the opportunistic fungi which can take hold and run rampant in patients with depressed immunological responses.[4] According to Trevor J. McGill, assistant professor of otolaryngology at Harvard University, this form of infection—fulminant aspergillosis of the nose and paranasal sinuses—is most likely to occur in children with acute leukemia who are undergoing chemotherapy and irradiation. Entering through the respiratory tract and attacking first the nose and paranasal sinuses, the fungus responsible for the disease can spread rapidly to the brain and lungs, in some cases causing death within two to four months.

Since this fungus is ubiquitous—appearing even in environmentally controlled clinical laboratories—it seems to be impossible to keep it from being inhaled by patients with lowered immunity. Early diagnosis, treatment with amphotericin B, and surgery seem so far to be the best means of combating the disease, but the Boston team of physicians headed by McGill reported that this was effective in preventing death in only about half of the cases they studied.

While incidents like these from widely dispersed regions were adding anecdotal evidence that the fungus diseases might warrant greater attention, several investigators at the Center for Disease Control headed by David W. Fraser were trying to get some measure of the national importance of these diseases. Their report, based on what they considered to be the best available figures, was published in the *Journal of the American Medical Association* late in 1979.[5]

The paper provided data on the incidence of the major systemic fungus diseases at 1,875 hospitals in all parts of the country which had reported to the Commission on Professional and Hospital Activities—about one-third of the nonfederal, short-term hospitals in the United States. This CPHA study, parallel to that done some years before by a team from CDC (see Chapter 1), compared data from the 1976 records with figures for 1970, the year reported in the previous paper.

Over the six-year period, cases of blastomycosis had decreased markedly (down 75 percent), and those of sporotrichosis slightly (down 7 percent), but in 1976 those two diseases accounted for only a minute fraction of the total—69 of the 4,274 cases reported. Increases were shown in all the other mycotic infections studied—4,205 of the 4,274 cases: candidiasis (up 9 percent), histoplasmosis (up 17 percent), coccidioidomycosis (up 74 percent), cryptococcosis (up 78 percent), actinomycosis (up 92 percent), and aspergillosis (up 158 percent). Factors given as contributing to the tremendous upward trend of coccidioidomycosis, cryptococcosis, and aspergillosis were the increasing number of patients being treated with immunosuppressive drugs, migration of susceptible persons into highly endemic areas, and aging of the population.

Histoplasmosis and coccidioidomycosis, diseases most often encountered in known endemic areas, were responsible for more than 75 percent of all cases of systemic fungus diseases reported by the participating hospitals in 1976. Aspergillosis, candidiasis, and cryptococcosis, generally classified as secondary infections in patients with compromised immunity and therefore not confined to specific geographic areas, showed by far the longest duration of hospitalization and the highest death rates.

The Fraser group at CDC also reported the cost of patient care for the 4,274 cases studied—a total of 57,948 days of hospitalization, which at the 1976 rate of $175 per day brought hospital charges alone to $10,140,900. They also projected these costs for the participating hospitals to all the nonfederal, short-term hospitals of the United States, arriving at a figure of more than $27 million in 1976 for hospitalization of patients with systemic fungus diseases. Adding physicians' fees, drugs, and other related charges would substantially increase this figure.

The 1976 data from the hospitals reporting to CPHA obviously provide only a part of the picture, since they do not cover afflicted persons who were not hospitalized, nor those in federal hospitals or chronic-care facilities—both of which probably care for substantial numbers of patients with systemic mycoses. The authors acknowledge these shortcomings, but believe the data are the best that can be obtained today, and probably the most accurate indicators of the incidence and morbidity of the systemic mycoses on a national level.

Perhaps the most helpful sign of greater attention's being paid to the fungus diseases is that while the report of the 1970 CPHA data appeared in *Sabouraudia*, a highly specialized journal of animal and human mycology with very limited distribution, the paper on the 1976 data was published by the *Journal of the American Medical Association,* which, with a circulation of nearly a quarter of a million, reaches most physicians in the United States.

This is no small advance, for physicians are the key figures in any attempt to get better data on the possible public health implications of the fungus diseases. There is no federal law requiring the reporting of fungus or any other diseases to the Center for Disease Control, and the individual states have their own laws as to which diseases will be reported within the state. Physicians are the ones who report diseases (where such reporting is required) to their respective state public health authorities. These officials, most of whom are also physicians, are the ones who report state figures to CDC, on what are classified as "notifiable" diseases, which the fungus diseases are not.

This information is lacking not just in the United States. Speaking before the Oholo Biological Conference in Maalot, Israel in 1976, Ajello of CDC said that no country in the world has made mycotic diseases notifiable to a public health agency. Without these vital statistics, he added, "support for medical mycological teaching, training and diagnostic centers, as well as basic and applied research, is difficult to justify and funding difficult to obtain from administrators. Medical mycologists must compete for support from a limited pool of funds against investigators of all other diseases. But the others, being notifiable, are backed up by data on morbidity and mortality that sway the minds of men and loosen purse strings."[6]

When sometime in the future such data become available in the United States to measure the true magnitude of the problem here, it is certain that the fungus diseases will not be found to rank with cancer, heart disease, or stroke as major threats. However, if complete national figures confirm the trends shown by the partial data in CDC's 1978 Annual Summary, cited in Chapter 1, and in the latest CPHA study, it should become evident that neglect has too long been visited upon the mycoses.

Reference Notes

Chapter 1. The Neglected Mycoses

1. Marie-Louise T. Johnson and Jean Roberts, *Skin Conditions and Related Need for Medical Care among Persons 1–74 Years,* Data from the National Health Survey, series 11, no. 212, Public Health Service, U.S. Department of Health, Education, and Welfare, November 1978.

2. Michael L. Furcolow, "Report on the Mycoses," National Institute of Allergy and Infectious Diseases, U.S. Department of Health, Education, and Welfare, 1976.

3. George E. Ehrlich, "Fungus Arthritis," *Journal of the American Medical Association* 240, no. 6 (August 11, 1978): 563.

4. Libero Ajello, "The Medical Mycological Iceberg," *Health Services and Mental Health Administration Health Reports* 86, no. 5, Public Health Service, U.S. Department of Health, Education, and Welfare (May 1971): pp. 437–448.

5. K. J. Hammerman, K. E. Powell, and F. E. Tosh, "The Incidence of Hospitalized Cases of Systemic Mycotic Infections," *Sabouraudia* 12 (1974): 33–45.

6. *Reported Morbidity and Mortality in the United States,* Annual Summary 1978, vol. 27, no. 54, Center for Disease Control, U.S. Department of Health, Education, and Welfare, September 1979.

Chapter 2. Microbiologist from Mississippi

1. Elizabeth Lee Hazen, "General and Local Immunity to Ricin," *Journal of Immunology* 13 (1927): 171–218.

2. Margarita Silva-Hutner, in an address to the Medical Mycological Society of New York at the Hazen memorial meeting in 1975.

Chapter 3. New England Chemist

1. Augustus B. Wadsworth and Rachel Brown, "A Specific Antigenic Carbohydrate of Type I Pneumococcus," *Journal of Immunology* 21 (1931): 245–253.

Chapter 4. A Unique Institution

1. *Standard Methods of the Division of Laboratories and Research of the New York State Department of Health,* 1st, 2d, 3d eds. (Baltimore, Md.: Williams & Wilkins, 1927, 1939, 1947).

2. Anna M. Sexton, *A Chronicle of the Division of Laboratories and Research, New York State Department of Health; the First Fifty Years—1914–1964* (Lunenburg, Vt.: The Stinehour Press, 1967).

3. Victor N. Tompkins, in Division of Laboratories and Research, *Annual Report,* 1962; Sexton, *A Chronicle.*

4. Walter Hollis Eddy and Gilbert Dalldorf, *The Chemical, Clinical, and Pathological Aspects of the Vitamin Deficiency Diseases,* 1st, 2d eds. (Baltimore, Md.: Williams & Wilkins, 1938, 1941).

Chapter 5. Search and Discovery

1. Elizabeth L. Hazen and Frank Curtis Reed, *Laboratory Identification of Pathogenic Fungi Simplified,* 1st, 2d eds. (Springfield, Ill.: Charles C Thomas, 1955, 1960); 3d ed. with Morris A. Gordon, 1970.

2. Albert Schatz and Elizabeth L. Hazen, "The Distribution of Soil Microorganisms Antagonistic to Fungi Pathogenic for Man," *Mycologia* 40 (1948): 461–477.

Chapter 6. Laboratory to Marketplace

1. Victor D. Newcomer et al., "The Valuation of Nystatin on the Course of Coccidioidomycosis in Mice," *Journal of Investigative Dermatology* 22 (May 1954): 431–440.

2. Rachel Brown and Elizabeth L. Hazen, "The Activation of Antifungal Extracts of Actinomycetes by Ultrafiltration through Gradocol Membranes," *Proceedings Society of Experimental Biology and Medicine* 71 (1949): 454–457.

Chapter 7. After the Invention

1. Edna Yost, *Women of Modern Science* (New York: Dodd, Mead, 1959), pp. 64–79.

Chapter 9. Concentration in Mycology

1. Howard W. Larsh, "The Public Health Importance of Histoplasmosis," in *Histoplasmosis: Proceedings of the Second National Conference,* edited by Libero Ajello, Ernest W. Chick, and Michael L. Furcolow (Springfield, Ill.: Charles C Thomas, 1971).

Chapter 11. Invention Repays Research

1. Norman F. Conant et al., *Manual of Clinical Mycology,* 3d ed. (Philadelphia: W. B. Saunders, 1971).
2. "Immunologic and Clinical Improvement of Progressive Coccidioidomycocis following Administration of Transfer Factor," *1974 Yearbook of Medicine* (Chicago: Yearbook Medical Publishers, 1974), p. 35.

Chapter 12. Diversity of Support

1. Carl W. Borgmann, *Twenty-five Years in Support of the Advancement of Science: An Evaluation of the Grants Programs of Research Corporation, 1945–1970* (New York: Research Corporation, 1972).
2. Sally Kelly, *Biochemical Methods in Medical Genetics* (Springfield, Ill.: Charles C Thomas, 1977).

Chapter 13. Nystatin in the News

1. Francis Rivett, "The Case of the Florentine Fungus," *Health News,* New York State Department of Health, 44 (May 1967): 18, 19.

Chapter 14. The Final Years

1. Barbara Sicherman and Carol Hurd, eds., *Notable American Women: The Modern Period* (Cambridge, Mass.: Harvard University Press, 1980).
2. Louis Haber, *Women Pioneers of Science* (New York: Harcourt Brace Jovanovich, 1979), pp. 63–73.
3. Rachel Brown, "Government," in *The Future Role of Women in Science and the World,* Special issue *The Chemist* 57, no. 1 (January 1980): 8.

4. Gilbert Dalldorf, "Self-Support of Medical Sciences through Patents," *Journal of the American Medical Association* 235, no. 1 (Jan. 5, 1976): 29, 30.

Chapter 15. A Continuing Challenge

1. American Society for Microbiology, *ASM News* 46, no. 1 (1980): 3, 4.
2. Neil M. Flynn et al., "An Unusual Outbreak of Windborne Coccidioidomycosis," *New England Journal of Medicine* 301, no. 7 (Aug. 16, 1979): 358–361.
3. David J. Drutz, "Urban Coccidioidomycosis and Histoplasmosis: Sacramento and Indianapolis," *New England Journal of Medicine* 301, no. 7 (Aug. 16, 1979): 381, 382.
4. "Rampant Fungal Infection Hits Nose First, Then Brain," *Wellcome Trends in Otolaryngology & Allergy* 1, no. 3 (August 1979): 3.
5. David W. Fraser et al., "Aspergillus and Other Systemic Mycoses: The Growing Problem," *Journal of the American Medical Association* 242, no. 15 (Oct. 12, 1979): 1631–1635.
6. Libero Ajello, "Systemic Mycoses in Modern Medicine," *Contribution to Microbiology and Immunology* 3 (Basel: S. Karger, 1977): 2–6.

Index

Actinomycetes, 75-77
Actinomycosis. *See* Fungus diseases: actinomycosis
Ajello, Libero, 30, 31, 199
Albany Medical College, 63, 70, 108, 114, 126, 127
Albert-Ludwigs-Universität, Pathologisches Institut (Germany), 68, 69, 115, 116
American Association of University Women, 188
American Chemical Society, 107
  Fisher Memorial Award, 115
American Institute of Chemists, 109, 189
  Chemical Pioneer Award, 109, 182
American Public Health Association, Albert Lasker Award, 116
American Society for Microbiology, 194
American Type Culture Collection (ATCC), 98, 104, 167, 168
Amherst College, Mass., 49, 50, 113
Amphotericin B, 27, 84, 124, 141-143, 159, 160, 197
Anderson, M.D., Hospital and Tumor Institute, Houston, 153, 154
Andreen, Brian H., 136
Antifungal preparations, 23-25, 27, 108, 142, 158. *See also* Amphotericin B; Nystatin
Antifungal therapy. *See* Fungus diseases: treatment of
Artis, William M., 156-158
Aschoff, Ludwig, 68
Aspergillosis. *See* Fungus diseases: aspergillosis
Aspergillus fumigatus. See Fungi: *Aspergillus fumigatus*
Aspergillus species. *See* Fungi: *Aspergillus* species
Athlete's foot. *See* Fungus diseases: athlete's foot

Barker, Joseph W., 95, 96
Barnard College, Columbia University, New York, 167
Barnhart, Frances. *See* Dalldorf, Frances
Bellevue Hospital, New York, 68, 119
Bender Hygienic Laboratory, Albany, 63, 64
Benham, Rhoda, 44, 45, 105, 109, 166, 184
  Award of the Medical Mycology Society of the Americas, 109
Berry, Janet E., 189
Biochemical Methods in Medical Genetics (Kelly), 166
Black, David G., Jr., 136
Blank, Fritz, 152-154
Blastomycosis. *See* Fungus diseases: blastomycosis

Borgmann, Carl W., 162
Bowdoin College, Brunswick, Maine, 115
Brown, Annie Fuller, 46-49, 53, 113, 114
Brown, George Hamilton, 46, 47
Brown-Hazen Committee. *See* Brown-Hazen grants program: advisory committee
Brown-Hazen Fund, 97, 98, 102
Brown-Hazen grants program:
advisory committee, 119, 123-125, 127, 129, 133, 134, 136, 169, 187
for biology departments, 123, 125-127
diversity of support by, 162-173
evolution of, 118-137
for medical mycology, 120, 132-150, 152-161, 166-168, 170-173
for Mexican institutions, 127-129
operation of, 118-137, 162-164, 190, 191
for research projects, 120-123, 140-144, 151-161, 171-173
for women in science, 129-131, 169, 170, 183
Brown-Hazen Lectures, 164, 165
Brown, Rachel Fuller:
education of, 46-53, 81
honors awarded to, 56, 86, 107-109, 188, 189
as member of Brown-Hazen Committee, 118-120, 129, 130, 133, 136, 167, 187
personal life of, 46-53, 112-115, 186, 188, 189
professional career of, 51-59, 74-80, 82, 98, 99, 104-109, 165
research on nystatin, 58, 59, 74-78, 84, 180, 181
research on pneumococcal polysaccharides, 52, 53, 55-57
research on tests for syphilis, 54, 55, 57
Scholarship/Fellowship for women in science, 169, 170, 187, 188
Brown, Ruth, 113, 189
Brown, Sumner Jerome, 46-49, 112, 113, 189
Bruner, Laman H., Jr., 189
Buchman, Edwin R., 97
Burt, Sarah. *See* Humphries, Sarah

Calderone, Richard A., 154, 155
California, University of, at Los Angeles, 194, 195
California, University of, Medical Center, San Diego, 159-161
Camp Mills, N.Y., 40
Camp Sheridan, Ala., 40
Candida albicans. See Fungi: *Candida albicans*
Candida species. *See* Fungi: *Candida* species
Candidiasis. *See* Fungus diseases: candidiasis
Capacidin, 108
Cardiolipin, 55, 57, 58, 80, 94
Carr, Emma Perry, 50
Catanzaro, Antonino, 158-161
Center for Disease Control (CDC), 25, 30-34, 193, 194, 197-200
Central High School, Springfield, Mass., 47, 48, 52
Ceratocystis ulmi. See Fungi: *Ceratocystis ulmi*
Chandler, Charles F., 63
Chemical Pioneer Award, American Institute of Chemists, 109, 182
Chen Sing-mei, 114, 188, 189
Chicago, University of, 50-53, 56
Chick, Ernest W., 147
Christ Church, Springfield, Mass., 47
Chromoblastomycosis. *See* Fungus diseases: chromoblastomycosis
Coccidioides immitis. See Fungi: *Coccidioides immitis*
Coccidioidomycosis. *See* Fungus diseases: coccidioidomycosis
Coccidioidomycosis Cooperative Treatment Group, 160
Coles, James S., 130, 169, 170, 183, 187, 188
Colorado College, Colorado Springs, 125
Columbia University, College of Physicians and Surgeons, New York:

Columbia University (*cont.*):
  affiliation of Elizabeth Hazen with, 40-45, 71, 105, 106, 108, 109, 183, 184
  as recipient of Brown-Hazen grants, 138, 164, 166, 167, 183, 184
Commission on Professional and Hospital Activities (CPHA), 32, 33, 197-200
Conant, Norman F., 147
  Manual of Clinical Mycology, 154
Consejo Nacional de Ciencia y Tecnologia (Mexico), 129
Cook Hospital, Fairmont, W. Va., 40
Cornell University Medical College, New York, 69
Costonis, Arthur C., 180
Cottrell, Frederick Gardner, 151, 161
Coxsackie viruses, 66, 69, 115
Cryptococcosis. *See* Fungus diseases: cryptococcosis
Cryptococcus neoformans. See Fungi: *Cryptococcus neoformans*
Cycloheximide, 77, 84

Dalldorf, Frances (née Barnhart), 68, 116, 191, 192
Dalldorf, Gilbert:
  as chairman of Brown-Hazen Committee, 119, 133, 136, 151, 162, 163, 187, 190, 191
  as director of Division of Laboratories and Research, 58, 59, 63-70, 79, 80, 86, 93-98, 115
  education of, 68, 69
  Fellowship in Medical Mycology, 91, 170-172
  honors awarded to, 115, 116
  memoirs of, 66, 191, 192
  personal life of, 68, 69, 116, 117, 191, 192
  professional career of, 68, 69, 115, 116
  research on virus diseases, 65, 66, 69, 116
  research on vitamin deficiency diseases, 69
Delbrück, Max, 122
Dermatomycoses. *See* Fungus diseases: dermatomycoses
Dermatophytes. *See* Fungi: dermatophytes
Dermatophytoses. *See* Fungus diseases: dermatophytoses
Dexter, Eugene A., 48
Dexter, Henrietta F., 48, 51
Dickey, Conway, 38, 112, 186
Distinguished Service Award, New York State Department of Health, 109
Distinguished Service Award, New York University College of Medicine, 115
Drutz, David J., 196
Dutch elm disease, 87, 88, 180, 181

Eddy, Walter H., 69
Ehrlich, George E., 30
Eisenhower, Dwight D., 175
Emory University, Atlanta, 155-158
Ewing, James, 69

Fisher Memorial Award, American Chemical Society, 115
Fleischmann Laboratories, Peekskill, N.Y., 51
Fleming, Alexander, 167
Florence (Italy) art works, restoration of, 89, 178, 179
Fraser, David W., 197-199
Fungal infections. *See* Fungus diseases
Fungi:
  dermatophytes, 24, 156-158, 167
  "opportunistic," 28, 29, 34, 168
  fungi discussed:
  Aspergillus fumigatus, 85
  Aspergillus species, 28, 29, 85
  Candida albicans, 28, 75-78, 85, 154, 155, 172, 191
  Candida species, 28, 29, 34, 85, 100, 101, 176, 182
  Ceratocystis ulmi, 87, 180, 181
  Coccidioides immitis, 158
  Cryptococcus neoformans, 75-78, 153, 154

Fungi (*cont.*):
fungi discussed:
Histoplasma capsulatum, 140, 141, 145, 154
Microsporum audouinii, 45, 73, 106
Sporothrix schenckii, 106
Fungicidin. *See* Nystatin
Fungizone, 124. *See also* Amphotericin B
Fungus diseases:
cost of treatment of, 24, 32, 156, 196, 198
diagnosis of, 26, 27, 29, 30, 71-73, 135, 145-150, 153, 154, 172
incidence of, 24-28, 44, 141, 145, 196-198
public health importance of, 24-26, 29-34, 193-200
reporting of, 25, 30-34, 196-200
treatment of, 27, 29, 99-102, 124, 141-143, 158-160
diseases discussed:
actinomycosis, 32-34, 198
aspergillosis, 27, 28, 32-34, 197, 198
athlete's foot, 24, 155, 156
blastomycosis, 26, 30, 32-34, 72, 194, 198
candidiasis, 27, 32-34, 72, 75, 100, 176, 198
chromoblastomycosis, 167
coccidioidomycosis, 26, 30, 32-34, 72, 150, 158-161, 196, 198
cryptococcosis, 26, 30, 32-34, 72, 75, 153, 198
dermatomycoses, 24
dermatophytoses, 155, 156
histoplasmosis, 26, 27, 30-34, 44, 72, 140, 141, 145, 150, 194, 196, 198
moniliasis (*see* candidiasis)
nocardiosis, 33
ringworm, 24, 156
sporotrichosis, 25, 30, 32, 33, 167, 198
tinea, 156
Furcolow, Michael L., 29, 33, 147, 148

Georgetown University, Washington, 154, 155
Goodman, Norman L., 148, 150
Gordon, Morris A., 169, 187
Laboratory Identification of Pathogenic Fungi Simplified, 169
Grappel, Sarah F., 153
Grasslands Hospital, Westchester County, N.Y., 66, 69
Guze, Samuel B., 140

Haber, Louis, *Women Pioneers of Science*, 188
Hamilton College, Clinton, N.Y., 123, 125
Harriman, W. Averell, 107
Harris, Annis (née Hazen), 36, 37, 185-187
Harvard University, School of Public Health, Cambridge, 138
Hazen, Annis. *See* Harris, Annis
Hazen, Clara, 38, 42, 111, 186, 187
Hazen, Elizabeth Lee:
affiliation with Columbia University, 40-45, 105, 106, 108, 184, 187
education of, 37-41, 43-45, 81
honors awarded to, 86, 90, 107, 109, 183, 187
Laboratory Identification of Pathogenic Fungi Simplified, 72, 105, 169
as member of Brown-Hazen Committee, 118-120, 129, 133, 134, 136, 167, 169
personal life of, 36-42, 71, 109-112, 185-187
professional career of, 40-45, 71-80, 82, 98, 99, 104-106, 108
research on *Microsporum audouinii*, 45, 73, 106
research on nystatin, 58, 59, 71-78, 84, 180, 181
research on ricin, 41
Scholarship for women in science, 169, 170
Hazen, Laura Crawford, 37
Hazen, Maggie Harper, 36
Hazen, Munson, 36
Hazen, Robert Henry, 37
Hazen, William Edgar, 36
Hazen, Willie Edgar, 36
Hearsey, Mildred, 111, 186
Hendry, Jessie, 114
Heuss, Theodor, 175
Hill, William C., 52

Hilleboe, Herman E., 119, 133
Hinkley, J. William, 95
Histoplasma capsulatum. *See* Fungi: *Histoplasma capsulatum*
Histoplasmosis. *See* Fungus diseases: histoplasmosis
Hobart and William Smith Colleges, Geneva, N.Y., 109, 177
Hopfer, Roy L., 153, 154
Hopkins, J. Gardner, 44, 184
Humphries, Sarah (née Burt), 41, 110
Hurd, Carol, ed., *Notable American Women: The Modern Period,* 187

Infectious Diseases Society of America (IDSA), 171
Ingraham, Hollis S., 133, 169
Instituto Politecnico Nacional (Mexico), 128, 129
Interferon, 69

Jensen, JoAnn S., 125, 126
Jones, Henry E., 155-158
Jong, Shung C., 168

Kelly, Sally, 165, 166
  Biochemical Methods in Medical Genetics, 166
Kentucky, University of, College of Medicine, Lexington, 138, 145-150
Kerkering, Thomas M., 91, 171, 172, 191
King, Kendall W., 133
Kirkbride, Franklin B., 94, 95
Kirkbride, Mary B., 94
Kobayashi, George S., 139-145

Laboratory Identification of Pathogenic Fungi Simplified (Hazen, Gordon, Reed), 72, 105, 169
Lamar, Lucius Quintus Cincinnatus, 38
Lamarr, Hedy, 177
Larsh, Howard W., 135, 147
Lasker, Albert, Award, American Public Health Association, 116
Lebanon Valley College, Annville, Pa., 125
Lederle Laboratories Division, American Cyanamid Co., 102
Lin Fei-ching, 114
Lowden, Inc., 180, 181
Lowden Tree Company. *See* Lowden, Inc.
Lyon, Mary, 50

McEwen, Currier, 68, 119, 133, 134, 169, 170, 172
McGill, Trevor J., 197
McGill University (Canada), 152
Maltaner, Elizabeth J., 54
Maltaner, Frank, 54
Manual of Clinical Mycology (Conant et al.), 154
Medical Mycological Society of New York, 187
Medical Mycology Society of the Americas, Rhoda Benham Award, 109
Medoff, Gerald, 139-145
Mexican institutions, Brown-Hazen support for, 127-129
Michalova, Dagmar, 188
Michigan, University of, Medical Center, Ann Arbor, 138, 155, 156
Microsporum audouinii. See Fungi: *Microsporum audouinii*
Mississippi Industrial Institute and College. *See* Mississippi University for Women
Mississippi University for Women, Columbus:
  education of Elizabeth Hazen at, 38, 39
  as recipient of Brown-Hazen grants, 131, 169, 170, 183
  as recipient of Hazen's books and papers, 112, 187
Moniliasis. *See* Fungus diseases: candidiasis
Mount Holyoke College, South Hadley, Mass.:
  education of Rachel Brown at, 48-51
  as recipient of Brown-Hazen grants, 130, 169, 170, 187, 188
Mount St. Vincent Nursing Center, Seattle, 185, 186
Mycoses. *See* Fungus diseases
Mycostatin, 99-101, 175. *See also* Nystatin

National Academy of Sciences, 79, 94, 115
National Conference on Histoplasmosis, Second, 31
National Foundation (March of Dimes), 115
National Health Survey, U.S. Public Health Service, 24, 156
National Institute of Allergy and Infectious Diseases (NIAID), 29, 33, 193-195
  Workshop on Medical Mycology Research and Training, 194, 195
National Institutes of Health (NIH), 155, 167, 171, 194, 195
National Science Foundation, 129, 167, 168
Nesset, Burton L., 125, 126
New York Academy of Medicine, 116
New York Academy of Sciences, 107
New York State Association of Public Health Laboratories, 61, 62, 105
New York State Board of Health, Sanitary Committee, 63
New York State Department of Health, Distinguished Service Award, 109
New York State Department of Health, Division of Laboratories and Research, Albany:
  early history of, 63, 64
  operation of, 54, 60-68, 105-107
  as recipient of Brown-Hazen grants, 120, 163-166
  research on antisera, 55-57, 64, 65
  research on fungus diseases, 44, 45, 65, 66, 71-80, 105, 106, 108
  research on tests for syphilis, 54, 55, 57
  research on virus diseases, 65, 66
New York State Hygienic Laboratory, Albany, 64
New York University, College of Medicine, 68, 119
  Distinguished Service Award, 115
Nocardiosis. *See* Fungus diseases: nocardiosis
Notable American Women: The Modern Period (Sicherman, Hurd, eds.), 187
Nourse, Walter B., 78, 83
Nystatin:
  commercial development of, 93, 94, 96, 98-101, 143
  issuance of patent on, 101, 174
  marketing of, 94, 99-102
  nonhuman uses of, 87-89, 178-182
  patenting and licensing of, 94-102
  research leading to discovery of, 58, 59, 71-78, 84, 180, 181
  royalties on, 102, 123, 132, 161, 168
  as a source of news, 174-183
  treatment of fungus diseases with, 29, 84, 99-102, 124, 143, 175, 176

"Opportunistic" fungi, 28, 29, 34, 168

Pacific Lutheran University, Tacoma, Wash., 125, 126
Pagano, Joseph, 98
Pangborn, Mary C., 55, 57, 58
Penicillium notatum, 167
Phalamycin, 108
Pneumococcal polysaccharides, 52, 53, 55, 56
Powers, Jack W., 136
Pyron, R. Scott, 136

Rake, Geoffrey, 93
Ramsey, Hal H., 127-129, 133, 136
Rapp, Fred, 165
Reed, Frank Curtis, *Laboratory Identification of Pathogenic Fungi Simplified*, 105
Research Corporation:
  as administrator of Brown-Hazen grants program, 118-131, 136, 190
  as administrator of nystatin invention, 95-103, 124, 151
  biological-medical sciences program (*see* Brown-Hazen grants program)
  departmental grants program, 123, 125-127
  nutritional sciences grants program, 97
  physical sciences grants program, 102, 121-123, 127, 162
  regional directors, 122, 127, 136, 147

Ringworm. *See* Fungus diseases: ringworm
Robinson, Lucena K., 53
Rosenblatt, Wilhelm F., 30

St. Peter's Episcopal Church, Albany, 53, 113, 114, 188, 189
Salisbury, David F., 182
Schatz, Albert, 73, 77
Schauer, Charles H., 119, 187
Schlessinger, David, 139
Seattle Pacific College, 125
Sexton, Anna M., 67, 115
Shadomy, H. Jean, 171, 191
Shadomy, Smith, 171, 172, 191
Shaw, Myrtle, 58
Sherman, Harry, 183
Shimer, Frances, School, Mount Carroll, Ill., 51, 52
Sicherman, Barbara, ed., *Notable American Women: The Modern Period*, 187
Sickles, Grace M., 66
Silva-Hutner, Margarita (née Silva), 45, 106, 109, 110, 166, 184, 187
Skin and Cancer Hospital, Temple University, Philadelphia, 152-154
Sloan-Kettering Institute for Cancer Research, Walker Laboratories, Rye, N.Y., 116
Smith, Sam C., 119, 121, 124, 133, 136, 140
Sporothrix schenckii. See Fungi: *Sporothrix schenckii*
Sporotrichosis. *See* Fungus diseases: sporotrichosis
Squibb Award in Chemotherapy, 86, 107, 176
Squibb & Sons, E. R., Division, Olin Mathieson Chemical Company:
  development of nystatin by, 93, 94, 96, 98-101, 143
  marketing of nystatin by, 94, 99-102
  Squibb Institute for Medical Research, 98
Standard Methods of the Division of Laboratories and Research of the New York State Department of Health, 60, 61
Stieglitz, Julius, 56
Streptomyces noursei, 78, 84, 98, 108, 167
Streptomyces species, 106, 108
Streptomycin, 75
Syphilis, tests for, 54, 55, 57

Temple University, Philadelphia:
  Health Sciences Center, 138
  Skin and Cancer Hospital, 152-154
Tinea. *See* Fungus diseases: tinea
Tompkins, Victor N., 67
Townsend, Francis J., 181
Townsley, Beulah, 111, 112, 186
Townsley, Paul, 186
Transfer factor, 159, 160
Transferrin, 157
Truman, David B., 187
Truman, Mrs. Harry S (Bess), 112

Universidad Nacional Autonoma de Mexico, 128, 129

Vassar College, Poughkeepsie, N.Y., 87, 130, 131
Veterans Administration Hospitals, 32, 33, 193, 194
Vietnam, fungus diseases among U.S. troops in, 24, 25
Virginia Commonwealth University, Medical College of Virginia, Richmond, 91, 171, 172, 191

Wadsworth, Augustus, 44, 54, 56, 61, 63, 65, 67, 68, 94
Wakerley, Dorothy, 53, 113, 114, 189
Waksman, Selman A., 75
Washington University, St. Louis, 138-145, 194
Waterman, Robert E., 97
Weber, Albrecht, 73, 74, 77
Westchester County, N.Y.:
  Department of Laboratories and Research, 69
  Grasslands Hospital, 65, 66, 69
Whalen, Robert P., 169
Wilbraham Academy, Wilbraham, Mass., 113
Williams, Robert R., 94, 97

Williams-Waterman grants program, 97
Women in the sciences, Brown-Hazen support for, 87, 129-131, 169, 170, 183
Women of Modern Science (Yost), 108
Women Pioneers of Science (Haber), 188
Workshop on Medical Mycology Research and Training, 194, 195

Yates, S. Blake, 95
Yost, Edna, *Women of Modern Science*, 108

The Fungus Fighters

Designed by Richard E. Rosenbaum.
Composed by Eastern Graphics
in 10 point Linotron 202 Palatino
with display lines in Palatino.
Printed offset by Thomson/Shore, Inc. on
Warren's Number 66 Antique Offset, 50 pound basis.
Bound by John H. Dekker & Sons, Inc.
in Holliston book cloth
and stamped in Kurz-Hastings foil.

Library of Congress Cataloging in Publication Data

Baldwin, Richard S., 1910-
The fungus fighters.

Includes bibliographical references and index.
1. Brown, Rachel, 1898-1980. 2. Hazen, Elizabeth Lee, 1888- . 3. Microbiologists—United States—Biography. 4. Chemists—United States—Biography. 5. Antibiotics—History. 6. Mycostatin—History. I. Title.
QR30.B34 615'.329'0922 [B] 80-69821
ISBN 0-8014-1355-9 AACR2